Training to Teach Adults
Mathematics

Acknowledgements

We would like to say thank you to David Kaye, Beth Kelly, Joan O'Hagan, Rachel Stone and the anonymous NIACE reviewer for taking time to read and comment on the various drafts of the text. We would like to thank all the trainee teachers we have worked with and learnt from.

A final thank you goes to Daian Marsh for keeping us grounded and reminding us of the importance of food as well as mathematics – and in particular those cakes on pi day (14 March).

Training to Teach Adults
Mathematics

Graham Griffiths and
Jackie Ashton, with
Brian Creese

Published by the National Institute of Adult Continuing Education
(England and Wales)
21 De Montfort Street
Leicester LE1 7GE

Company registration no. 2603322
Charity registration no. 1002775

NIACE is the National Institute of Adult Continuing Education, the national voice for lifelong learning. We are an internationally respected development organisation and think-tank, working on issues central to the economic renewal of the UK, particularly in the political economy, education and learning, public policy and regeneration fields.

www.niace.org.uk

For details of all our publications, visit http://shop.niace.org.uk

Follow NIACE on Twitter: @NIACEhq
@NIACECymru (Wales)
@NIACEbooks (Publications)

Cataloguing in Publications Data
A CIP record for this title is available from the British Library

978-1-86201-846-4 (Print)
978-1-86201-847-1 (PDF)
978-1-86201-848-8 (ePub)
978-1-86201-849-5 (Kindle)

All websites referenced in this book were correct and accessible at the time of going to press.

The views expressed in this publication are not necessarily endorsed by the publisher.

Produced by Full Stop Communications Ltd, www.fullstopcommunications.co.uk

Cover and text design and typesetting by Patrick Armstrong Book Production Services
Printed in the UK by Charlesworth Press

Contents

Introduction

This book is an introduction to teaching mathematics and numeracy in the post-compulsory sector.[1] Many people have a good understanding of what is meant by the school and university sectors, but the sector that this book is aiming at is not so straightforward.

The sector includes further education colleges, adult education institutions, community based learning, and a variety of independent (usually work-based) training providers. The learners involved will either be young people aged 14–19 or adults aged 19 and upwards. Some of these learners will be studying with a particular career in mind, while others will be looking to pass a mathematics qualification in order to move into further study. Some may wish to succeed in a subject that they have previously had difficulties with and some may study out of interest. Some adults want to help their children with school mathematics. The courses may lead to a range of qualifications including the current school-based GCSE and GCE qualifications, those intending to support mathematics within employment and everyday activities (at the moment labelled 'functional skills') and small 'bite size' units intended for adults. More qualifications, such as 'Core mathematics' for those who have GCSE grade C but are not going on to study formal mathematics, are currently being developed. Many of these courses are across a range of levels from Entry Level 1 (considered as Entry Level in the National Qualification Framework [NVQ]) through to advanced-level skills (NVQ Level 3).

It is likely that you will end up working in one or two areas of this potentially bewildering array of options, and it may help to consider which of these options most interests you. Wherever you finally settle to teach, it is worth bearing in mind that mathematics and numeracy teaching presents a challenge across all age groups, in all settings and at all levels. It is fairly well established that many people find it difficult to grasp the ideas that are studied and, while the subject of mathematics is engaging for some learners, there are many who feel the subject is not for them. Mathematics features in the news headlines – usually through the problems that occur. There are concerns about the position of one country in relation to another, concerns about 'standards' over time and concerns about errors made in everyday activities. All of these are well known to be problematic areas and there is no clear consensus of how to move forward. There are a range of views and influences on teaching mathematics that will come from researchers in education, inspection regimes and government. Institutions will make choices about the qualifications they choose to use and have expectations of their teachers. It is also worthwhile remembering the reasons why people undertake mathematics courses after they finish compulsory schooling. Mathematics is an academic subject that can be studied in its own

---

1. The term 'post compulsory' is used in this book to describe the post-school, non-higher education sector which covers further education colleges, adult education centres, training providers, community education, and prison education, involving a range of learners from age 14 upwards. We recognise that education is now compulsory for individuals up to age 18 (albeit that this includes training within employment). Other potential terms have their own difficulties: further education excludes the wide range of settings mentioned in the text, adult education excludes the 16–18 age group, lifelong learning is supposed to mean from cradle to grave, education and training (the current name for the sector) could be a description of any part of education and may not last as a term. We have therefore chosen to use 'post compulsory'.

right; is a service subject for other academic subjects such as science and economics; can help in an employment situation; and can help in everyday lives. These are not mutually exclusive categories, nor are they intended to be exhaustive, but the categories may help us to consider a range of issues as we plan provision.

It is therefore important for the professional working in mathematics education to be open to the debates and to seriously consider the range of approaches to teaching, whatever their views. This text is intended to assist a teacher of mathematics and numeracy at the start of their journey.

Some readers may already be teachers of other subjects who may wish to move into teaching mathematics for some or all of the time. Some of the book will contain ideas that such readers will have been introduced to in other, more generic, teacher training programmes. In this text, we have made an attempt to avoid replicating standards texts and focus on illustrating ideas with examples drawn from mathematics and numeracy practice.

There is likely to be at least a few years between your own experience of being taught mathematics and your first foray into teaching mathematics or numeracy. You may have been taught mathematics in this country or another country but no matter when or where you were educated, your experience of learning mathematics is likely to differ considerably from how mathematics is currently taught to adults. Many trainee teachers report that they were taught mathematics in what would be termed quite a traditional manner, often with the teacher standing at the front of the class and very little interaction between the students. This type of teacher-led teaching, where the teacher talks and the students listen, is known as the transmission teaching style. With this approach, the teacher introduces the topic, possibly explaining associated terminology, talks through the steps involved in any calculations, gives one or two examples, possibly asking some questions to check for understanding of the processes involved, and then asks the students to work on some exercises individually so that they can practise the skills he or she has just 'transmitted' to them. The skills are generally taught in a fairly abstract way, with a focus on learning the rules and getting the 'right' answer, and there tends to be little relation to real life or any attempt at explaining why the skills and knowledge are necessary.

On the other hand, current adult mathematics and numeracy teachers are likely to use different approaches to this transmission style. Learners are likely to be more actively involved in their learning, both in deciding what they would like to learn and in the actual learning and assessment process. There may be a mixture of whole-group work, small-group or pair work and individual work. Learners may be given activities to carry out and discuss in small groups. There may be whole-group activities with discussions and learners may be working with a range of learning materials including ICT materials, practical apparatus and realia, in addition to the more traditional worksheets and textbooks. Teachers also try to make the learning relevant to the learners and link what they are learning to real life. Following changes in the curriculum, there is now a greater emphasis on the process of solving problems rather than just getting the right answer.

At the end of this book you will find links to some key curriculum documents and sites which you will need to work with as a specialist teacher. The book has been written to avoid linking too strongly with any particular curriculum in order to allow the ideas to stand the test of time. Of course, we understand that it is important for new teachers to become familiar with the curricula they use – this will come with the development of practice in real classrooms.

The learners (Chapter 1) introduces the reader to the range of learners they might expect to be working with through case studies and learners' own words. It avoids using a deficit model of the learners, focusing instead on the resources they bring to a learning situation and how to build on their previous experience. It considers the variety of goals that learners might have and explore ways of discussing and negotiating the learning curriculum with them.

Mathematical knowledge (Chapter 2) lays the foundations of knowledge about mathematical thinking that a teacher is expected to be aware of. The chapter focuses on developing the subject knowledge of the teacher rather than outlining what is taught to learners. This includes an understanding of how mathematical concepts have developed over time in order to count and measure. It includes how mathematics has developed into a problem-solving tool and the stages involved when working through problems. It also takes a look at understanding data including how statistics can be used for different purposes. The chapter looks at the use of mathematics in a range of contexts and issues of how mathematics is best developed.

Educational theories and approaches to teaching mathematics and numeracy (Chapter 3) looks at theory and practice. It looks at key learning theories and considers where they fit in relation to teaching and learning mathematics. The chapter then outlines a number of approaches to teaching that are important to mathematics and numeracy teaching, including investigative approaches, collaborative approaches, multicultural and ethno-mathematical approaches, and critical mathematical education.

The teaching and learning cycle (Chapter 4) builds on some of the information in previous chapters. It covers the range of activities that effective teaching depends upon, such as diagnostic assessment (for identifying learner needs), planning lessons focusing on mathematics and numeracy, and checking learning and assessing learner progress.

Teaching practice (Chapter 5) takes the new teacher through their first steps in the classroom, considering the learning environment and the beginnings of effective classroom management. It includes some discussion of how current teaching approaches may be different from their own experiences at school. It explores the different forms of teaching practice (working with individuals and groups; training classes and group teaching practice) and how these might be affected by context. You will be given pointers about how to make full use of the learning opportunities offered by teaching practice and how to work effectively with a mentor.

Examples of teaching activities (Chapter 6) takes on board the previous chapters and gives examples of activities that are suggested as good practice. The chapter offers a range of ways of seeing the choice of activity, through the choice of context and setting (and hence a mixture of curriculum areas), as a problem-solving tool, and also through curriculum focus such as number, measure, shape and data handling.

Planning for inclusive practice (Chapter 7) explores a number of issues around inclusion that a teacher needs to take into account when planning for teaching mathematics and numeracy. It focuses on how to make a programme of learning relevant and interesting for the learners and how to build positively on diversity in the classroom. It will also discuss how a teacher can make their teaching accessible to all learners.

Finally, **Professionalism (Chapter 8)** explores the term 'professional' in relation to the mathematics and numeracy teacher. Readers will be invited to interrogate the term and consider what it means in relation to their own role as a mathematics or numeracy teacher. The chapter also addresses what happens when the new teacher comes to the end of their initial training. It looks at what is expected of an effective mathematics and numeracy teacher and how they can evidence this to observers (such as Ofsted inspectors or managers). It invites new teachers to consider what they have learned already and how to address any gaps that they perceive in their knowledge and skills.

We hope that by reading this book you will gain some insights into the real world of teaching in the post-compulsory sector. The final chapter identifies some key characteristics of a teacher of mathematics in the sector and we invite you to consider your strengths and weaknesses against these characteristics and how best to develop your practice as a teacher.

1 The learners

INTRODUCTION

Learners of mathematics or numeracy in the post-compulsory sector can be any age from 14 upwards. They are from many different backgrounds in terms of their ethnicity and culture; some have English as their first language but many do not. They have different past educational experiences and levels of attainment and are motivated to attend mathematics/numeracy classes for a variety of reasons. However, in order to teach the learners effectively we, as teachers, must get to know them and relevant aspects of their backgrounds very well. Below is a sample of six learners who attend different mathematics/numeracy classes. The contexts represented include GCSE mathematics, Level 1 and 2 functional mathematics, family numeracy and mathematics for vocational learners. The case studies contain information about each learner, their motivations for studying mathematics and their learning goals, their strengths and the resources they bring with them, as well as particular areas of difficulty. There is also a discussion of the strategies used to support them and how these are selected and negotiated with the learners.

The samples only scratch the surface in terms of representing the many different types of learners who attend post-compulsory mathematics or numeracy classes. Learners have different backgrounds contributing many different factors that can affect the way that they view mathematics and how they are able to access the teaching and learning in the classes that they attend. Here are some of the factors that may have an effect:

> **Gender**: Although in recent years GCSE results have been broadly similar for females and males, traditionally mathematics has been seen as a male subject. It has been argued that certain types of activity and assessment have favoured different genders, for example the idea that coursework works in favour of females. Alternatively, others argue that where there are preferences this is due to different expectations being placed on learners by society, through parents and teachers.

> **Age:** Very young learners have more recent memories of being taught mathematics and may remember methods more clearly and have better study skills. This can make them appear more confident. On the other hand, they may feel under pressure from peers not to conform in a classroom setting. Some younger learners may may not be ready to distinguish adult learning from school education, which may affect the way they behave in class and relate to the teacher. Older learners have more experience of life that can be utilised, but may not have studied for a while and therefore may feel that they have forgotten a lot of what they were taught. This can lead to a lack of confidence in a learning situation.

> **Previous educational experience and attainment**: Mathematics is a subject that many people have strong feelings about and this may be linked to positive or negative experiences at school. Some adults have very low confidence in mathematics and may

suffer from a condition termed 'maths anxiety' because they found mathematics lessons traumatic. They may have been humiliated because they couldn't keep up with others in the class or were unable to give correct responses to questions from the teacher. Some adults even experienced physical punishment when unable to do what the teacher asked, such as reciting times tables correctly. Some simply disliked the way that the subject was taught and could not see the relevance of it. These negative experiences often lead to people finding learning mathematics difficult at school and therefore they may not have achieved a qualification in the subject, or achieved low grades.

❱ **Language:** Some adults have low-level English language skills, perhaps because English is not their first language. Some may have studied mathematics to a reasonably high level in their first language.

❱ **Literacy levels:** Some learners will have very good levels of literacy, which will obviously make learning in general easier and will support them in learning mathematics. Others will have lower levels of literacy, which means that they will have difficulty with reading and comprehension of written material. This will be a barrier to learning mathematics or numeracy and succeeding in written assessments.

❱ **Ethnicity and cultural background:** This may affect the way that mathematics was taught, the value placed on mathematics education, family pressure and expectations of achievement. Cultural background can have positive or negative influences on the way adults view and learn mathematics.

❱ **Preferred ways of learning:** Learners respond differently to different teaching and learning strategies. This can affect their motivation to learn and also how well they absorb and retain information. There are debates over whether individuals have a definable 'learning style', and some are concerned that labelling individuals may itself create a difference in performance, but there is a consensus that individuals can have preferences in relation to teaching and learning activities.

❱ **Learning difficulties:** Some learning difficulties can create barriers to learning unless the teacher is aware of them and is skilled in adapting teaching and learning strategies to enable full access to the learning material. Examples of learning difficulties include dyslexia, dyspraxia, dyscalculia, Asperger syndrome and attention deficit hyperactivity disorder (ADHD). As with most factors that affect learning, the teacher should be aware of the strengths that the learners have that can be emphasised and drawn upon.

❱ **Motivations for studying mathematics:** If a learner has a strong drive to succeed in something that requires them to improve their mathematics or pass a mathematics qualification, this can be strong enough to overcome many existing barriers to learning the subject.

❱ **Employment background (previous and current) and other interests:** Adults have a lot of experience that should be drawn upon and referred to when they are learning mathematics. Often the person themselves does not realise that they have been using mathematics and this may need highlighting. Interests and hobbies can also provide a rich source of learning contexts for problem solving.

TASK 1.1

(a) How might you group the factors above under broader headings?

(b) Think about ways in which the following may affect learning:

   (i)   What the learner feels about their previous educational experiences

   (ii) The extent to which the learner uses mathematics/numeracy outside the classroom

   (iii) Whether the learner has dyslexia or any other learning difficulty

   (iv) Learner views of how they feel they learn best

   (v) Whether the learner came to this country from a war zone

   (vi) Whether the learner lives alone or with family or friends

   (vii) Whether the learner has access to the internet

   (viii) The types of leisure activities the learner does

   (ix) Whether the learner is employed

   (x) Whether the learner had any education beyond primary level

   (xi) Whether the learner has children

   (xii) Whether the learner was bullied at school

Suggested responses to tasks can be found at the end of the book.

CASE STUDIES

In the case studies below, look out for references to any of the factors noted and consider how they may be affecting the learning of mathematics/numeracy. It is worth remembering that the case studies are learners' accounts of their past and while these clearly include some facts (for example, where they studied), some aspects will be perceptions (for example, there is evidence to suggest that learners are more likely to blame themselves than their institutions of learning for their difficulties in progressing).

CASE STUDY 1

S is a 15-year-old white British female. She was born and educated in the UK and her first language is English. She left school early at the age of 15 with no qualifications and enrolled on an art and design course at Level 1. Her interests lie in art and fashion and she would like to be either a fashion designer or an architect when she is older.

She disliked mathematics at school reporting that she couldn't understand her mathematics teacher and felt she was not progressing. Now, she has been required to enrol on a mathematics course in conjunction with her art and design course rather than opting in to study. She can see the value in being able to do mathematics, however, as she recognises that she uses it in her everyday life, for example to prevent 'getting ripped off' in the shops. She also acknowledges that most jobs require you to understand mathematics and to have a mathematics qualification. She couldn't think of a single job that didn't require mathematics.

The learners on the course were assessed in mathematics and placed in a mixed-level functional mathematics class (Entry Level 2 to Level 1) designed specifically for art and fashion students. This learner was broadly assessed as being at Entry Level 2, with extra support required in number. The course's focus on the vocational subject meant the course was of interest to the learner, and promoted a desire to succeed in that subject along with an acknowledgement of the importance of mathematics.[2]

The results of the assessments were shared with the learners and discussed in tutorials. The discussion allowed the learner to say what they think they find difficult and whether they agree with the results. This learner reported that she had particular problems with percentages and fractions, which was consistent with the results of the assessment. The learners were also encouraged to talk about their feelings about mathematics, how they have felt about it in the past and if they have any preferences for how they like to learn – for example, if they enjoy practical work or working in groups. The teacher takes the opportunity to discuss why mathematics is important and how it is relevant to the particular student's short- and longer-term goals.

With this particular learner, the teacher found that she was actually very capable in mathematics but could be easily distracted. Therefore the key was to keep her engaged by trying to link to her interests as much as possible and to use practical teaching and learning strategies that played to her strengths. After a period of study the learner still has trouble understanding mathematics but it makes a bit more sense to her now that it is linked to art. For example, she feels she can understand decimals a little now.

Summary of factors potentially affecting S's learning of mathematics

) *Previous educational experience:* This appears to have had a significant negative effect as the learner disliked mathematics at school because she couldn't understand the way that it was taught. She left school early without any qualifications, which may have affected her confidence and self-esteem in a classroom setting.

) *Age*: The class consists of a very young group of learners which may affect the way that they behave in class and ultimately affect an individual's concentration. This could be why the learner is easily distracted.

2 The teacher is currently undertaking a subject specialist teacher training course in numeracy to develop his expertise in teaching mathematics. He also has a passion for art and design and is combining his skills to teach mathematics to art and fashion students. He can therefore plan a course of study that aims to teach mathematical topics through the context of art and design wherever possible; for example, teaching symmetry by looking at symmetry in art and by creating symmetrical patterns.

❱ *Interests*: Linking mathematics to her interest in art seems to be having a positive effect, as she can see the relevance of it.

❱ *Motivation for learning mathematics*: This appears to be having a positive effect because the learner needs to succeed in mathematics in order to achieve her goal of working in art and fashion or architecture.

CASE STUDY 2

L is a white British female in her mid-thirties. She was born and educated in London and her first language is English. She left school at 17 with a few GCSEs including grade D in Mathematics. She never really enjoyed mathematics at school because she found it difficult and confusing. She considered herself to be not very good at the subject. However, she now has two young children and has embarked on several courses of study for various reasons. L decided she would like to work as a teaching assistant, specialising in supporting children with special needs. She embarked on a teaching assistant course and realised she would also need to improve her mathematics skills in order to support the school children with their work. As a result, she also joined a family numeracy class to better understand mathematics and to find out how children learn mathematics so that she could be more effective as a teaching assistant and could support her own children with their learning. At the same time, she joined a GCSE Mathematics class in order to improve her skills and get a better grade. She is working part time as a teaching assistant in a primary school to gain some experience of working with children and can see the benefit of having good mathematics skills and knowledge of current teaching methods.

Having ascertained that the learners were prepared for it, as it was an open access mixed-level course the tutor gave the learners a written assessment in the first session in order to get an idea of their levels. Results were shared and discussed with individuals and particular areas of difficulty were discussed. The teacher also asked the learners what they hoped to achieve by being on the course and how they could work together to achieve those targets. As it was a short (ten-week) course, the teacher suggested that the learner should identify two mathematical areas that they most wanted to work towards and these were recorded on an individual learning plan. This learner stated that her problem areas were fractions and percentages, so it was agreed that she would be supported to improve in those areas.

The learner brought a high degree of self-motivation and determination to succeed in the range of learning situations. As she was assessed to be the most able learner on the course and was also getting support through attending a GCSE class, the teacher was able to agree that she would work independently on her targets some of the time. It was also agreed that she would support another learner in the group in paired activities at times during the course, as this would not only help the other learner but would also help L to develop skills that would be useful in her role as a teaching assistant.

Summary of factors potentially affecting L's learning of mathematics

❱ *Previous educational experience*: L did not enjoy mathematics at school, which has affected her confidence as she felt she was not good at it.

❯ *Age*: In her mid-thirties, this learner has a fair amount of life experience and is able to see the benefit of having good mathematics skills.

❯ *Motivations*: This learner is highly motivated by a strong desire to succeed in her chosen career as a teaching assistant and to support her children's education. This enables her to work independently as well to seek support from different sources, such as the various courses she is studying.

CASE STUDY 3

A is a black British male aged 15. He attended school in the UK but has been identified as having attention deficit hyperactivity disorder (ADHD) and had to leave school early due to the linked behavioural difficulties. He was required to enrol on English and Mathematics GCSE at a local community college because of his age and was therefore an unwilling class member. He seemed to have a very negative view of education and could be angry at times. He could not see the point in having to study mathematics and felt that his level of knowledge and skill was sufficient for life and work. His biggest love was music and he also enjoyed drama and had an interest in sport.

The teacher gave the learners an assessment in the first session of the course so that she could identify their strengths and areas for development. The assessment consisted of a written test and self-assessment through a set of activities on different topics from the GCSE syllabus. The results were discussed with A and three key areas for development agreed and recorded on an individual learning plan. The self-assessment activity allowed the teacher to discuss the learner's response to different teaching and learning strategies, such as small-group work using games and tactile materials. The teacher was able to assess that A was happy to work in pairs or small groups and was interested in the different visual and tactile materials presented. She ascertained that he was very good at mental arithmetic but disliked mathematics because of having to record his methods on paper, which he found challenging. He also disliked the areas that required a lot of written work, such as drawing graphs or charts. However, his written work was fairly good in other subjects that he enjoyed. The teacher was able to explain to him that a high proportion of marks were allocated in the Mathematics GCSE for demonstrating the processes used to obtain the answer (the workings out). He was quite unhappy about that but agreed that he would try to record his workings out as long as he was given a decent break in the middle of the session.

A therefore brought a certain confidence in his ability to work things out in his head. He was unafraid of topics such as algebra, which some of the group found very challenging. He liked to be actively involved in discussions and liked to come to the board to show what he could do. His ADHD gave him an energy which could sometimes be injected into class activities to create a positive environment. The challenge was to keep him engaged and to keep discussions on topic as he liked to 'play the clown' and would often try to distract other learners in the group. To assist with this, an agreement was made that he could listen to music while working on written exercises. The teacher also aimed to include topics that would be of interest, such as working out averages in the context of sport, and to build on his strengths. For example, A was very capable of solving problems that involved money. The teacher also encouraged A to explain or demonstrate how he had obtained answers to others in the group before suggesting he wrote down the steps on paper.

Summary of factors potentially affecting A's learning of mathematics

) *Previous educational experience*: A was made to leave school early without any qualifications. As a consequence, he had a very negative experience of school and teachers generally, which affected his willingness to participate in class.

) *Learning difficulty*: ADHD affected his previous educational experience and attainment to an enormous degree and made it difficult for A to concentrate in his current class. However, he did have energy which could be used in a positive way.

) *Age*: A was young and this appears to have affected the way he viewed the learning experience – as something 'done' to him rather than through self-motivation. Nevertheless, he was confident in his mathematical abilities, which gave the teacher opportunities to value his skills and involve him in group activities.

) *Literacy*: The learner's literacy level is good, which may be supporting his mathematics learning. However, he dislikes writing in relation to mathematics which makes it more challenging for him to succeed in written assessments.

) *Motivation for learning mathematics*: He was forced to attend the class and therefore unwilling to participate. He also had trouble seeing the relevance of the learning material as he felt his mathematics was already good enough. However, he was able to recognise that he did need to work on recording his methods in order to gain sufficient marks in the GCSE exam.

) *Interests*: The teacher was able to use his interests in music and sport to engage and motivate him.

CASE STUDY 4

T is a white British male in his early twenties. He has lived in the UK all of his life and was educated here. He left school at 16 with few GCSEs, gaining a grade E in Mathematics. He did not enjoy mathematics at school, mainly because he did not feel confident in it. As a consequence he felt he would not succeed in the subject and was easily distracted in class. He did not have a clear idea of what he wanted to do after leaving school and therefore did not consider that he would need to pass mathematics for his future.

He studied a course in vehicle maintenance after leaving school and felt his mathematics was sufficiently good to cope with the demands of the course; consequently he chose not to do any further study in mathematics at that time. He also worked part time stacking shelves in a large hardware store where he felt he did not require any mathematics at all in order to do his job.

However, he started to develop an interest in a career in the sports industry; he had previously enjoyed and pursued various sports as a leisure interest but had not considered making a living out of it. He decided to enrol on a Level 1 Diploma in Sport and Active Leisure course as he became passionate about a career as a personal trainer. He also opted to study mathematics again alongside the course as he felt that he would require a good level of mathematics in his chosen career. For example, he was conscious of the need to calculate body mass index. He enrolled on a Level 1 Functional Mathematics course and he feels that the course is also beneficial to aspects of his everyday life, such as money calculations. Now that he can see the benefit of the mathematics course he has relaxed and is enjoying it; he no longer gets distracted so easily in class and feels he is making progress. He is a little older than many of the other class members and can see them behaving in a similar way to the way he did in mathematics classes at school. He has started to talk to them about how important it is to try to concentrate on the lesson and he feels they are more receptive to his encouragement than to similar messages from the teacher.

Summary of factors potentially affecting T's learning of mathematics

) *Educational experience*: T did not experience any success in mathematics, which affected his confidence. This, in turn, resulted in him having very low expectations of success in the subject which, coupled with his lack of enjoyment, meant that his concentration in class was poor and he was easily distracted.

) *Motivation for learning mathematics*: He could not see the relevance of the subject while at school. He had no real career aims at the time and, therefore, nothing to work towards. Now he has decided upon a career he can see the relevance of learning mathematics and has opted to study the subject.

) *Age*: When he was younger he was more easily distracted in class; now he is older he is getting frustrated by the behaviour of the younger group members who are behaving in a similar way to himself when he was their age. They seem to be responding well to someone a few years older than themselves, encouraging them to concentrate.

) *Interests*: T was very interested in sport while he was at school but no reference was made to anything that interested him in mathematics lessons at that time. Now he is using his own interest in sport as motivation to study mathematics further.

CASE STUDY 5

R is an Asian female in her mid-thirties. She was born and educated in Bangladesh but has been living in the UK for more than ten years. Her first language is Bengali but she has a fairly good level of spoken and written English. She is a full-time mother of three children but would like to train to be a mathematics teacher when her children are older. She liked mathematics at school and as an adult but felt her mathematics education had not been good and that she needed to improve her English. As a result, she decided to enrol on both mathematics and English courses at her local adult and community college. An assessment before joining the course showed she needed support at Level 2 for mathematics and English and she enrolled on courses for both.

Once she joined the mathematics course, she was asked to complete a diagnostic assessment online. This provided some feedback directly to R and gave detailed feedback to the teacher on which topics she was doing well in and which she required support in. The results showed R was quite capable in number but needed support with data handling. The teacher spoke to R about the assessment – how she found it and what she felt that it signified. During discussion with the teacher, R agreed that she did need help with statistics and probability and that it was probably due to the fact that she was not taught data handling in school. The teacher also discovered that it may also be related to the language demands of this area of mathematics as English was not her first language.

The teacher found that R was quite confident about performing written calculations. She also found that R loves to help her children with their mathematics homework and feels she is just managing to stay ahead of them. Her enthusiasm for learning in general and mathematics in particular, together with her determination to become a mathematics teacher were felt to be strengths that could be harnessed, not solely to support R but for the whole group. It was discovered that R had received a fairly traditional, transmission style of teaching and was not used to working in pairs or small groups. The teacher explained to R that she intended to use a very different approach to how she had been taught because it was found to be more effective and enjoyable. Despite feeling that her mathematics education was not good, R was initially sceptical and was reluctant to work with others. However, once she started to become friendly with other group members, she relaxed and started to participate willingly in paired and small-group activities. She discovered that she enjoyed explaining things to other learners and they, in turn, were able to explain certain things to her such as contexts or terminology that she was unfamiliar with.

The teacher discovered that language was a barrier for this learner, which was interfering with her ability to understand the functional skills problems. This was tackled by looking at a range of written problems and unpicking them, in addition to focusing on the terminology associated with different mathematical topics. Learners were also encouraged to write their own mathematics questions on different topics and whole-group discussions were held on topics such as budgeting for meals.

The teacher also discovered that while the learner was very good at arithmetic calculations, she tended to do them very quickly and, as a result, sometimes made mistakes. The teacher decided to tackle this in several ways – she shared different calculation methods with the group which demonstrated that there was not a single correct method that had to be used for performing calculations. She discussed why the different methods worked, to encourage the learners to think about what they were doing rather than doing it robotically. She also taught checking methods and encouraged learners to use them to check their answers.

Summary of factors potentially affecting R's learning of mathematics

❱ *Language*: English is not this learner's first language and appears to be preventing her reaching her full potential, especially when dealing with written problems. However, R is being supported with her English language and literacy skills by attending an English course.

❱ *Previous educational experience*: R enjoyed mathematics as a child, which helped to give her a very positive attitude towards it, but felt her education had not been good, which also seems to have driven her to want to improve. She had been taught in a very traditional way, which initially hampered the teacher's ability to get her to work with others. There had been a focus on arithmetic where she was educated, which gave her confidence in number and calculation; however, there had been less coverage of areas such as data handling, meaning that she needed support in this area. She also feels the need to work very quickly, which leads to her making errors; this may be linked to previous educational experiences.

❱ *Cultural background*: R has a very positive view of the importance of education and a love of mathematics in particular.

❱ *Motivation*: R is very keen to improve her own skills in order to be able to help her children and achieve her long-term goal of becoming a mathematics teacher.

CASE STUDY 6

G is a black African female in her early thirties. She was born in Ghana and had her early education there, but has lived in this country for about fifteen years. Her first language is Twi but her English speaking and writing skills are extremely good. She has a foundation degree and is currently working as a teaching assistant in a school. She would like to eventually run her own nursery school and realises that she needs a good level of mathematics in order to progress in her chosen career. She is currently studying Level 1 Mathematics at her local adult and community college and hopes to go on to achieve Level 2 next year. She didn't enjoy mathematics at school at all; the teaching methods at her school were very traditional and the teachers at the school were extremely strict. She quite likes mathematics now; however, she lacks confidence in her ability.

As with R above, G completed an online diagnostic assessment. This indicated that she was stronger in the area of number than in measure, shape and space, or data handling. In discussions about the diagnostic assessment with the teacher, G stated that she felt she needed support in all three areas of mathematics and did not feel confident in number despite the results of the assessment, but agreed that measure was her particular bugbear that she could not 'get her head around'. What this learner also brought to the class was a friendly, open nature that would help to gel the group and promote collaborative learning. She joined the class with a friend who was working at a higher level and they were found to work well together in paired activities. She was also open to working with anyone else in the group and was keen to participate in whole-class discussions. G was also keen to improve and was determined to conquer what she saw as her weak areas.

Looking at the assessment results, it was agreed that her key targets would be to convert between units of measure within the metric system, to perform calculations involving time and to calculate with decimals. The teacher also deduced that confidence was the key barrier for G and that she should focus on building her confidence by planning for success through small achievable learning objectives. Apart from an interest in working with children, the learner expressed an interest in cooking. The teacher planned to use both of these interests to create familiar contexts for the learner, which should help with confidence and achievement. For example, when looking at calculation methods, the teacher asked learners which different methods they were familiar with; G was able to talk about current school methods she was aware of which made her realise that she had some knowledge that other group members did not possess. When teaching measure, the teacher related weight and capacity to the context of cooking. G measured by eye rather than weighing or measuring ingredients accurately using measuring instruments, but during discussion realised that she must have had a sense of the measurements required in order to do this effectively.

Summary of factors potentially affecting G's learning of mathematics

❱ *Previous educational experience*: G did not enjoy learning mathematics as a child as the teachers were very strict and this may have affected her confidence in her ability.

❱ *Language and literacy*: English is not her first language but she seems to have a good level of English skills and has succeeded in achieving a foundation degree. Completing the degree may have provided her with study skills that are now assisting her ability to learn and it also indicates that she has good literacy skills, which should also support her learning of mathematics.

❱ *Work and interests*: G is working in education, which has given her up-to-date knowledge of a range of calculation methods. This, in turn, has boosted her confidence. She is interested in cooking, which has provided the teacher with a relevant teaching context.

❱ *Motivations*: G is motivated to succeed in mathematics by her desire to run her own nursery.

CASE STUDY 7

M is a black British female in her late twenties. She was born and educated in the UK. M hated mathematics at school; she left school at 18 with two A-levels but grade G in Mathematics GCSE. She retained her negative attitude towards the subject and had very little confidence in her ability. She has a young child and joined a family numeracy class at the local children's centre because she was worried about transferring her negativity to her child. Her nine-year-old niece asked her for help with her mathematics homework and

M was embarrassed that she wasn't able to help her.

Family learning is a route into learning that can draw in adult learners who might be reluctant to enrol on more formal adult mathematics courses. It is quite likely that family learning tutors might encounter learners with poor experiences of learning mathematics, who have very negative feelings towards the subject but have been motivated to overcome their fears by the opportunity to help their children. Because of this, the teacher discussed reasons for joining the course with the learners in the first session and ascertained that M had very strong feelings about mathematics and was quite anxious about it. The teacher decided to offer the choice of using the written assessment in an alternative way for those who preferred to do so – as a 'can do' assessment. Using this method, learners would look at the questions and identify which ones they could do and which would be difficult for them, rather than actually working through them.

This learner felt that all areas of mathematics were problematic but stated that she particularly struggled with multiplication, division and decimals. From the discussion that took place as part of the assessment, the teacher concluded that M was operating at Entry Level 3. The key target areas were recorded on an individual learning plan and the teacher also probed M's reasons for disliking mathematics so intensely. It was discovered that she felt that mathematics was taught badly in a style that didn't suit her, whereas other subjects were taught in more interesting ways. She realised that mathematics was important in life but had never experienced it taught in a way that showed the relevance of everyday mathematics. The subjects she had enjoyed were taught in a more interactive, practical way but mathematics had been taught with an emphasis on remembering 'the rules', which she felt she would never be able to do. The teacher realised that the key to supporting M was in giving her the confidence to help her child and in showing her that mathematics could be enjoyable. She explained that M would be shown the methods that are used in school and that games and activities would be used for learning.

M found that she enjoyed learning the school methods as they explained what was happening and simplified the learning material for her so that she could make sense of it. Once she started to feel she could understand some of the concepts, she relaxed and started to enjoy learning. She reported that the practical teaching approaches suited her learning style and that she found the games and activities fun. The teacher ensured that she made links to everyday mathematics wherever possible, for example by talking about best buys when looking at percentages and M appreciated this. M also found the focus on mathematics vocabulary useful for both her own understanding and for communicating with her child and niece. She said that she felt she had been equipped with the tools for helping her child and found she had developed enough confidence to help her niece with her homework. She became an active participant in group discussions and brought a creative approach to the group by making suggestions for activities that could be used at home to help children with mathematics. She also demonstrated her determination to continue to develop her own skills by buying a mathematics textbook and working on some exercises at home. Evidence that her confidence had grown came when she asked to be entered for a mathematics test at the end of the course – she passed a functional mathematics qualification at Level 1.

> **Summary of factors potentially affecting M's learning of mathematics**
>
> ❯ *Previous educational experience*: M had very strong negative feelings towards mathematics and it had badly affected her attitude towards and confidence in the subject.
>
> ❯ *Motivations*: M was highly motivated by wanting to support her child and niece with their mathematics. This enabled her to overcome her fears and join a class in the first place. The teacher was able to adopt the methods used in school to teach her and improve her confidence.
>
> ❯ *Preferred ways of learning*: M felt that she had been taught in a style that did not suit her and much preferred the practical teaching approaches used in the family numeracy class. She also liked to see the relevance of what she was learning.

SUMMARY

The case studies represent a small sample of the different types of learners that attend adult mathematics/numeracy classes. Some of the factors that affect learning have been discussed but there are many others that teachers should be aware of, for example dyslexia. Different motivations for accessing mathematics learning have also been discussed but, again, other motivations exist, such as access to higher education or improving the chance of gaining employment. What the case studies demonstrate is how essential it is to get to know the learners and why they are in your class. To help the learner to achieve their goals it is important to identify their strengths and tap into them, in addition to diagnosing areas that require improvement and identifying any barriers to learning. It is essential to involve the learner as much as possible in deciding on the most appropriate strategies to be used. You should also find out what experience and interests they have so that you can use familiar contexts that will motivate and engage them.

FURTHER READING

Griffiths G. and Stone R. (eds) (2013) *Teaching Adult Numeracy: Principles and Practice.* Maidenhead: Open University Press.

Swain, J., Baker, E., Holder, D., Newmarch, B. and Coben, D. (2005) *'Beyond the Daily Application': Making Numeracy Teaching Meaningful to Adult Learners.* London: NRDC.

2 Mathematical knowledge

INTRODUCTION

This chapter will look at the subject matter of this book. You will explore mathematical ideas which are intended to develop your own knowledge of the subject. You might take some of these ideas into your classes with learners or you might not; you will find ideas that will translate into classroom practice, for example the tasks on different ways to calculate, but other ideas may be for your own development only. For example, it is helpful for a teacher to consider working with different numeral or number systems to experience learning for themselves, though you may choose not to use these ideas with your learners.

You may call the subject mathematics or numeracy, or some other term. Some writers consider that numeracy is a subset of mathematics, the aspects of mathematics that involve number or perhaps what you need for everyday life. Others think that numeracy is larger than mathematics and really means the application of mathematics to the real world.

Look at the following tasks and decide whether you think they are mathematics tasks, numeracy tasks or both. The remainder of the chapter will consider questions such as the following and discuss the subject matter issues raised.

    a. What is 15×680?

    b. Calculate 20% of £1500.

    c. What is the sum of angles in a polygon with n sides?

    d. How far is the Sun from the Earth? (image and lengths)

    e. What is the chance that it will rain tomorrow?

    f. Which group has performed best? (graphs needed)

    g. How many babies were born in the UK last year?

    h. How many different ways are there of arranging n cubes?

    i. How do you decide what product to buy in a supermarket?

    j. When do you need to set off for work in the morning?

    k. The items in a shopping basket total £4.75. What is the change from £20?

COUNTING AND MEASURING

One view of mathematics is seeing the subject as working with counting and measuring. Certainly historically one can see the use of number to count (and account) and measure amounts of tradable goods, such as the number of sheep or an amount of grain. Some of the earliest evidence of human civilisation consists of objects with markings that suggest counting and recording. Such recording led to a variety of numeral systems, one of the most well-known examples being the *Roman system*. Even today, television programmes sometimes have the year of production given in Roman numerals; for example, MMXIII would be 2013. Instead of just using one stroke for each unit (like a tally system), groups of strokes are collected with new letters so that, for example, IIIII = V and VV = X.

Consider: What is MMXIII – C? What is MMXIII × CVII?

While the Roman system was useful for recording, there are some problems with this type of system. It was not used for calculating – the Romans used boards with pebbles to carry out calculations – and as numbers get bigger more symbols are needed (what is a billion in Roman numerals?).

Other systems, called *positional systems*, developed a way of using a limited set of symbols to represent all numbers using place value. Such a system is the one we use today (called the denary system) where ten symbols – 0, 1, 2, 3, 4, 5, 6, 7, 8, 9 – can be used to represent all whole numbers. For example, 243 means two hundreds, four tens and three units, whereas 324 means three hundreds, two tens and four units. This system is sometimes known as the *Hindu–Arabic system*, which acknowledges the Indian and Middle Eastern sources of mathematical knowledge. Indeed, it is worth noting that European scholars were reluctant to reject their use of Roman numerals for many years.

There are other positional systems, such as the *binary system* which is useful within computing. The binary system uses the digits 0 and 1 only and instead of using 10 as a base for the arithmetic, uses 2. In binary, the number 'two' is represented by 10 – i.e. one lot of two plus no units. In binary 11 represents one lot of two plus one unit – what we would write as 3 in denary. The denary number 4 is 2 squared, and would represent the next position in the binary system, i.e. 100 (one lot of 2^2, no 2s and no units). The next binary position would be for $2^3 = 8$, so 1000 in binary is 8 in denary, 1001 in binary would be 9 in denary, 1010 in binary would be 10 in denary and so on.

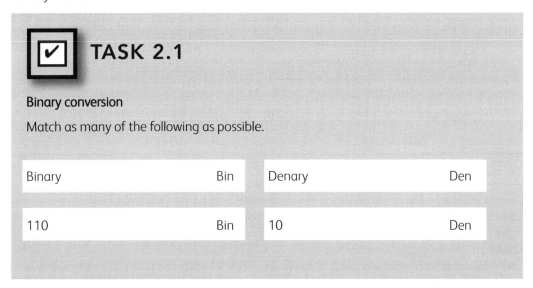

☑ TASK 2.1

Binary conversion

Match as many of the following as possible.

Binary	Bin	Denary	Den
110	Bin	10	Den

11011	Bin		33	Den
1011	Bin		6	Den
1000	Bin		27	Den
100010	Bin		8	Den
1010	Bin		34	Den
100110	Bin		12	Den

Suggested responses to tasks can be found at the end of the book.

(This task is based upon a similar activity produced for the Skills for Life Quality Initiative Level 4 professional development modules for adult numeracy teachers.)

CALCULATIONS IN BINARY

Similarly to arithmetic in our normal system, it is possible to calculate within the binary system. There are simpler rules for number bonds (additions) and multiplication than for denary as there are fewer digits involved. Nevertheless, for some individuals working with binary can be a disorienting experience. For someone in that situation, working with binary arithmetic may give them a better understand how a learner can struggle with normal, denary arithmetic.

Number bonds

+	0	1
0	0	1
1	1	10

Multiplication table

+	0	1
0	0	0
1	0	1

TASK 2.2

Complete the following arithmetic problems in the binary system (without converting back to denary). Use the same algorithms that you use for ordinary arithmetic. What information did you need in order to complete the arithmetic in the binary system?

101011 + 111

1011 − 110

111011 × 1101

11011 ÷ 1001

Suggested responses to tasks can be found at the end of the book.

(This task is based upon a similar activity produced for the Skills for Life Quality Initiative Level 4 professional development modules for adult numeracy teachers.)

ZERO

It is worth noting that the use of positional number systems requires the use of *zero*. The Roman system had no need for a symbol for nothing – what was the point? But positional systems need something as a placeholder in numbers such as one hundred and eight (i.e. there are no tens). The ancient Babylonians used a symbol as a placeholder in their semi place value system, although it was not fully used as a number in its own right. The notion of a zero is now attributed to Indian mathematicians for which there is documentary evidence originating from the ninth century.

MEASURING

Aside from counting, the other main function of numbers is in *measuring*. What was the last thing that you measured? You may have measured a space in your home to fit a cupboard, or weighed out some flour in a recipe. Perhaps you used a machine which measured for you, such as the pumps at garages which measure the petrol in litres, or the scales in supermarkets which weigh and price vegetables. Alternatively, you may not have needed to be precise and used something like paces to measure a distance – or even a walking time ('it's about 10 minutes' walk') – or about half a pack of sultanas. Clearly there are times when accuracy is needed and times when it is not.

 TASK 2.3

What examples can you think of where accuracy in measurement is needed and where it is not?

Suggested responses to tasks can be found at the end of the book.

CALCULATING

In adult mathematics or numeracy classes, there will inevitably be learners who were taught different methods of calculation, particularly 'long' multiplication and division. There are many ways to calculate, and teachers need to be careful not to suggest any one method is the 'correct' one. We encourage all teachers to try out many different approaches to what are often called the *four operations* of addition, subtraction, multiplication and division. There is also a range of terminology about such calculations that it is important for a teacher to know.

Number bonds are numbers that add up to 10, such as 2 and 8, 3 and 7.

Bridging uses the nearest 10 as a part of a calculation, for example 17 + 5 is the same as 17 + 3 + 2 = 22. This is using number bonds to get to the next 10, i.e. 17 + 3 = 20, and then recognising that to finish off the original sum requires 2 more.

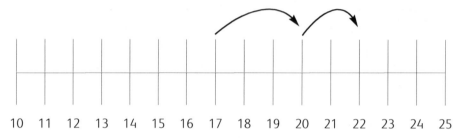

Near doubles involves noticing that some additions are close to doubling (which most people find straightforward), for example 19 + 18 is double 18 plus 1, i.e. 36 + 1 = 37.

Partitioning involves breaking down a number into its component parts and dealing with these separately. For example, in 382 + 116, we have 300 + 80 + 2 and 100 + 10 + 6. Adding these separately gives 400 + 90 + 8 = 498.

3	8	2	+	1	1	6				
3	0	0		1	0	0		4	0	0
	8	0			1	0			9	0
		2				6				8
							=	4	9	8

Compensation is where a simpler problem is calculated and the answer adjusted. For example, 23 + 19 is 23 + 20 minus 1; i.e. 43 − 1 = 42.

Multiplication facts/times tables, such as 2 × 6 = 12, 7 × 8 = 56, and so on.

X	1	2	3	4	5	6	7	8	9	10
1	1	2	3	4	5	6	7	8	9	10
2	2	4	6	8	10	12	14	16	18	20
3	3	6	9	12	15	18	21	24	27	30
4	4	8	12	16	20	24	28	32	36	40
5	5	10	15	20	25	30	35	40	45	50
6	6	12	18	24	30	36	42	48	54	60
7	7	14	21	28	35	42	49	56	63	70
8	8	16	24	32	40	48	56	64	72	80
9	9	18	27	36	45	54	63	72	81	90
10	10	20	30	40	50	60	70	80	90	100

Repeated addition – multiplication can be seen as repeated addition: 19 × 3 = 19 + 19 + 19 = 20 + 20 + 20 − 3 (compensating) = 60 − 3 = 57.

Repeated subtraction is dividing by counting how many times you subtract to get to zero.

15 − 5 = 10, 10 − 5 = 5, 5 − 5 = 0, so 15 ÷ 5 = 3.

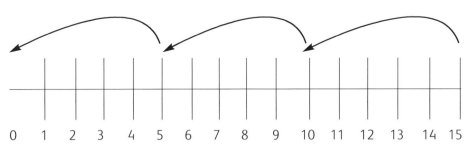

0 1 2 3 4 5 6 7 8 9 10 11 12 13 14 15

Chunking is a useful way of dividing by using repeated subtraction but in 'chunks'. For example, for 180 ÷ 5, I note that 10 lots of 5 is 50. So I can take away chunks of ten 5s.

180 − 50 = 130, 130 − 50 = 80, 80 − 50 = 30, so there are 30 lots of 5 so far.

30 − 5 = 25, 25 − 5 = 20, 20 − 5 = 15, 15 − 5 = 10, 10 − 5 = 5, 5 − 5 = 0.

Altogether we have 30 plus 6 lots of 5 in 180.

In other words, 180 ÷ 5 = 36.

TASK 2.4

The following tables A–D contain a number of fairly basic calculations involving small numbers. Try out the calculations – do what comes naturally to you. Then after completing the questions, consider where you used the same methods and where you used different ones.

A) How would you add these numbers?	B) How would you subtract these numbers?	C) How would you multiply these numbers?	D) How would you divide these numbers?
2 + 8	10 − 7	3 × 4	12 ÷ 3
4 + 8	14 − 8	8 × 4	24 ÷ 6
12 + 10	33 − 10	12 × 10	70 ÷ 10
17 + 15	23 − 14	12 × 15	90 ÷ 15
18 + 19	75 − 23	18 × 19	90 ÷ 6
28 + 46	54 − 6	28 × 23	240 ÷ 12
125 + 8	54 − 19		
145 + 19			

Suggested responses to tasks can be found at the end of the book.

(This task is based upon a similar activity produced for the Skills for Life Quality Initiative Level 4 professional development modules for adult numeracy teachers.)

WORKING WITH SHAPE

A significant element of mathematics through history has been the study of shape, usually described as geometry. Many cultures have left us evidence of graphical representations; one well known for dealing with geometry is the ancient Greek civilisation. They used geometry to understand more about number rather than explicitly deal with fractions and they used ratios of sides of geometric objects to understand non-integer quantities. They also used geometry to develop the notions of what we now call *proof*. In the famous *Euclid's Elements*, from some explicitly stated starting points much of what is now studied as geometry in secondary schools (and beyond) was developed.

Take the notion that the three angles in a triangle add up to 180°.

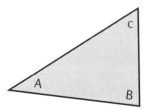

Consider the following arguments. Which one do you believe in or trust the most?

Argument 1

I measure lots of angles in a triangle to the nearest degree and they all added up to 180°. So all the angles in a triangle add up to 180°.

Argument 2

If I tear off the corners of a triangle they all line up on a straight line. Therefore the angles all add up to 180°.

Argument 3

Take any triangle; add a parallel line as shown. Angle A and A' are the same, C and C' are also the same. And as A', B and C' make a straight line then A' + B + C' add up to 180° and therefore so do A, B and C.

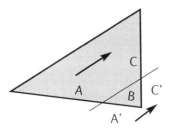

The first argument is an *induction argument*. It is certainly quite effective in suggesting that the angles add up to 180° but we might not be sure we are measuring accurately and that there may be a funny triangle that we haven't thought of which doesn't add up to 180°. The second argument is a development of the first but also hints that there something about bringing the angles together and is a movement toward argument 3. Argument 3 is a more formal *deduction argument* that is intended to demonstrate for all cases that the angles add up to 180°. It does require some knowledge to be established first – for example that A and A' are the same.

WORKING WITH DATA

A later development in the mathematical sciences is the study of probability and statistics. In the seventeenth century French mathematicians Pascal and Fermat corresponded over games of chance. The discussions they had still influence the types of questions studied today.

Consider the following question. Which is more likely, getting a head and tail when tossing two coins or getting two heads?

The discussions between Pascal and Fermat raised the key issue that we still discuss today. There are three possible outcomes when two coins are tossed (two heads, two tails and one of each) but this does not mean a third of a chance each. The one instance of each face can actually happen in two different ways (HT and TH) which makes it twice as likely as getting either two heads or two tails.

First coin → ↓ Second coin	H	T
H	HH	TH
T	HT	TT

Dealing with such compound, controlled situations is the basis of mathematical probability. From such starting points more complex situations can be considered.

TASK 2.5

Put the following statements in order of likelihood.

a. When I buy a lottery ticket, I win the jackpot.

b. It will rain tomorrow.

c. The next plane I take lands safely.

d. The chance of recovering from a diagnosed breast cancer.

e. One Direction has another number one single.

f. The National Health Service is privatised.

Suggested responses to tasks can be found at the end of the book.

How do we associate probabilities to such events? (a) involves a more complex version of the coin-tossing example but produces a fairly clear answer, whereas (b)–(d) require the collection of past data which are then interpreted using probability theory to produce answers for which there are various degrees of confidence in the answers. Essentially, any answers to (e) and (f) are almost certainly guesswork. These are the main ways that probability is assigned when applying probability theory – through the *classical probability* approach (a), the *empirical probability* approach (b–d) or the *subjective*, best-estimate approach of (e) and (f).

Statistics involves the collection of empirical data which is then used either to summarise an existing situation or, more likely, to predict future events. For statisticians, this is not a subset of mathematics but an entirely different area. Some statisticians have argued that mathematicians' focus on processes and applications of particular tools has distorted the need to understand the context and the problem itself.

If you search for a definition of statistics you are likely to see the following words – or similar ones – associated with *data*.

❱ Collection or organisation

❱ Analysis

❱ Presentation or representation

❱ Interpretation

Many traditional mathematics curricula and assessments, certainly those up to and including Level 2/GCSE, have focused on such activities as:

❱ completing data collection tools such as tally charts and frequency tables;

❱ drawing or reading from such bar charts, pie charts, scatter diagrams or similar; and

❱ calculating values of measures of location such as mean, mode and median and measures of spread like the interquartile range.

Most of these activities have a single answer in mind – sometimes perhaps an approximation within a range of values. It is these activities that some statisticians think misrepresent statistics. They are concerned that more open-ended discussions about how and why samples are chosen and the limitations of studies are missed out.

 TASK 2.6

In groups choose one of the following situations and answer the question. Be ready to explain the situation and your answer to the whole group.

Situation 1

A newspaper editor wishes to form an opinion of the public's reaction to a matter discussed in Parliament, and invites readers of his paper to write to him expressing their views. Why does this produce a bad sample?

Situation 2

Some of the children at a school in a large town are engaged in a project which, in part, is concerned with the newspapers purchased and read by the town's population. In order to obtain a sample representative of the town they decide to ask every child in the school to report which newspapers are regularly purchased by his or her family. Assuming a total response, comment on the possible existence of bias.

Situation 3

2000 telephone subscribers resident in a particular parliamentary constituency are selected at random from a telephone directory and are asked how they intend to vote at the next general election. This sample is intended to be representative of the whole population of this constituency. Why might this sample be biased?

Suggested responses to tasks can be found at the end of the book.

(This task is based upon a similar activity produced for the Skills for Life Quality Initiative Level 4 professional development modules for adult numeracy teachers.)

PROBLEM SOLVING

Rather than focusing on topics, more recent thinking on mathematics is concerned with the process of undertaking mathematics. The Hungarian mathematician George Pólya wrote a number of books during the twentieth century which investigated various aspects of *mathematical problem solving*.

But what is meant by mathematical problem solving? Although we could see questions such as 'what is ¾ of 240?' as a mathematical problem, it is not what is really meant by a problem in Pólya's writing. Pólya uses the following problem to illustrate his thinking: find the diagonal of a cuboid of which the length, the width and the height are known. (Actually Pólya uses the term rectangular parallelepiped rather than the more usual cuboid that we would normally use.) This problem is assuming that the learners are familiar with Pythagoras' Theorem and many of the problems included in the book *How To Solve It* are set at higher GCSE level and early Advanced level.

There are other types of problems that could be investigated at other levels. There are the investigation-type tasks that have been proposed over the years:

What is the relationship between the edges, vertices and faces of three dimensional shapes?

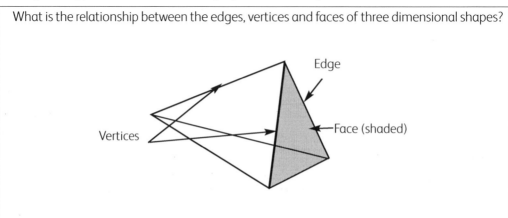

How many handshakes would there be between a given number of people in a room?

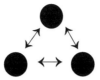

What is the maximum volume of an open cuboid made by cutting squares out of the four corners of a rectangle and folding up the sides?

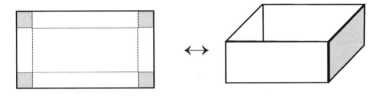

These problems can be investigated by a range of learners with the patterns spotted being expressed either in words or using algebra. Undertaking such extended tasks with learners may be seen as representing more authentically what mathematics means to mathematicians. For many, mathematics is not about applying a short set of rules to come up with 'answers' but rather a more extensive, problem-solving activity.

The Hungarian mathematician Pólya wrote extensively about problem solving. He argued that problems are solved by going through four stages:

a. Understand the problem

b. Devise a plan

c. Carry out the plan

d. Look back on your work.

These stages can be applied to the above investigations – but they could also be used when looking at the type of 'word problem' often used in mathematics curricula. These four stages and problem-solving approaches influenced the Cockcroft Report of the 1980s and the introduction of GCSE Mathematics and then later qualifications such as key skills and functional skills. Both key and functional skills have used the three process headings of: *representing*, *analysing* and *interpreting* which are not the same as Pólya's four stages, although there are links. Representing incorporates 'understanding the problem' and to some extent 'devise a plan'; analysing also involves 'devise a plan' along with 'carry out the plan'; and interpreting is much the same as 'looking back'.

While functional mathematics includes a focus on process, the problems it uses are less the abstract problems of Pólya and investigations and are intended to be more useful in everyday life – hence 'functional'. A typical type of problem would be to consider a family, information about various holidays and a budget and ask for appropriate choices to be made.

A different sort of problem which has been found useful in mathematics education is attributed to the Italian Enrico Fermi, who set various estimation problems. These *Fermi problems* give little information and expect participants to make reasonable estimates to provide an approximate solution.

> An example: how many babies were born in the UK in one year?

This problem can be resolved in a number of ways. For example, we could assume that there are roughly the same amount of people of any given age within the UK, babies are the ones that are 0 years old. If we assume a top end of around 80 years, then dividing the total population of the UK, let us say 60 million, by 80 should give us our estimate of about 750,000.

Another calculation might be to estimate the number of women of child-bearing age and then estimate the proportion that might conceive in a given year. So you might say that half the population – 30 million – are female, that half of them are of child bearing age – 15 million – and that 1 in 20 of these will have a child in a given year. This also gives an estimate of 750,000.

It is worth noting that these are reasonable estimates. For example, in 2012 the Office for National Statistics recorded 812,970 births.

MATHEMATICS IN SHOPPING

An important study into the mathematics that people use in their everyday lives was undertaken by Jean Lave and others. This study involved following shoppers around a supermarket and asking them to explain their decision-making processes. What was found was that individuals used a range of information and tools to make their decisions – only one of which is mathematics. For example, when looking at different quantities of the same product, it is not necessarily the 'best value' size that gets chosen (as in maximising the amount per unit cost) but rather a combination of this mixed in with storage space available at home and the likelihood of wastage.

You have probably seen questions like the following in resources and on test papers.

Which of the following bottles of oil is the best value?

  a. 75 cl costs £3.50

  b. 1 litre costs £4.00

The expected answer is that bottle b is the 'best value'. The work of Lave and others suggest that in such a *situated* activity the answer is more complex than this. By engaging the everyday, the teacher may actually then bring to the fore issues that were not intended. For example, individuals may choose products because they fit in their cupboards or they may consider that buying a larger product will encourage overconsumption or waste.

MATHEMATICS AT WORK

The amount of research being conducted on workplace mathematics has increased over the years. There has been work on street sellers in Brazil, a number of studies into nurses (many in the UK) and some work into a variety of other occupations.

Similarly to the shopping study mentioned above, these studies often note the way in which mathematical activity is *situated* in other activities which lead to answers that might not be expected in the mathematics classroom.

An interesting example is taken from the study into street sellers. When selling coconuts, the street sellers tended to sell them up to three at a time. This meant that they were very clear on the prices of 1, 2 and 3 lots of coconuts. So when they were asked what 10 coconuts cost, they quickly worked out 3 lots of 3 plus 1, rather than some calculation involving changing the decimal place.

Another issue involved in mathematics at work is the required accuracy of any response. In engineering, there will be times when a high degree of accuracy will be required, whereas in other areas there may be less need for accuracy. For example, in one nurses study where drug doses were being regulated and pressures from other parts of the job meant that following the rules was not straightforward, the nurses had different ideas about how the drug doses would work – all of which may be acceptable within the limits expected.

THE PURPOSE OF MATHEMATICS AND NUMERACY EDUCATION

So what is the purpose of mathematics education?

For some it is the development of mathematics as a subject in its own right. This ever-expanding subject can still offer treasures of knowledge, although often in some strange esoteric world such as the solution to Fermat's Last Theorem. It is worth bearing in mind that the twentieth century mathematician G.H. Hardy argued that he had never done anything useful in his life – meaning that his mathematics was entirely abstract and would have no application. As it turns out, much of the work he was involved in was significant in the development of encryption processes that we use in computers and mobile devices today. Even so, such applications are not easily understood and will be studied and developed by those who take forward mathematical sciences to a higher level. This set of people is not a majority and this purpose is unlikely to be suitable for all. Yet mathematics is a subject that all study.

What other purposes are there? There is a case for understanding the place of mathematics in human history. The earlier sections outlined some of the developments from a historical perspective, noting their importance in the economy and culture. Many have argued that it is important that individuals have enough mathematics knowledge to be able to run their lives in an appropriate way. Others suggest that we need everyone to have enough mathematics knowledge for our economy to develop in an increasingly technological world. Others want individuals to develop critical thinking and consider that problem solving is an important life skill.

SUMMARY

We should return to the notions of mathematics and numeracy. Through all of the developments mentioned above there are mathematical activities that seem to be internal problems and others that explicitly look at the connection to the real world. For many, the problems that connect to the real world are the defining feature of what might be termed numeracy. The more abstract, mathematically internal problems would not be considered numeracy. This is not necessarily related to level – or 'the basics' – but about application. Whatever term you choose to use, it is important for you as a mathematics teacher to have clear idea of what you think mathematics and numeracy are and how they relate to various curricula that you will teach.

FURTHER READING

Griffiths, G. and Stone, R. (eds) (2013) *Teaching Adult Numeracy: Principles and Practice.* Maidenhead: Open University Press.

Ma, L. (2010) *Knowing and Teaching Elementary Mathematics: Teachers' Understanding of Fundamental Mathematics in China and the United States*, Studies in Mathematical Thinking and Learning Series (2nd edn). London: Routledge.

Pólya, G. (1990) *How to Solve it: A New Aspect of Mathematical Method.* London: Penguin.

3 Education theories and approaches to teaching mathematics and numeracy

INTRODUCTION

The field of education has developed over the years and studies have provided a range of perspectives towards understanding how people learn – and, indeed, what is to be learned. It is important to note that understanding learning is not the same as understanding teaching. Some theories of learning say little about how to teach, while others have more obvious connections. In many cases, understanding a theory gives a way of interpreting and talking about teaching and learning rather than suggesting ways of practising.

This chapter will present some key education theories and consider some examples of practice in mathematics education that link to them. There are theories on what constitutes a curriculum as well as how people learn it. We will first take a look at Bloom's taxonomy of learning domains before turning to categories of learning theories and how these theories influenced what we might mean by curriculum, along with examples drawn from mathematics education. Learning theories are generally classified into three main types: behaviourist, cognitivist and humanist. Behaviourists are first in the development of ideas, followed by cognitivists and then humanists. These are broad headings that are being used and in this introductory text such ideas and examples are not intended to be full descriptions of the range of ideas related to these theoretical perspectives. Instead the examples are intended to illustrate some key practical issues that trainee mathematics/numeracy teachers are likely to encounter and link to theoretical perspectives.

We will then explore a number of approaches to teaching that are important to mathematics and numeracy teaching, namely: investigative approaches, collaborative approaches, multicultural and ethnomathematical approaches, and critical mathematical approaches. The reason for identifying these approaches is not so that you choose one of these but rather that you can understand how approaches are developed to emphasise certain aspects of learning mathematics. Indeed, it is important to understand that there are not clear distinctions between these approaches and that some activities you choose to use might be categorised under more than one of these. In addition, not all individuals who make a claim to a particular approach will do the same things – these approaches are really broad headers for those with similar (but not identical) views and values.

CATEGORISING THE CONTENT OF SUBJECTS

NOW, what I want is, Facts. Teach ... nothing but Facts. Facts alone are wanted in life. Plant nothing else, and root out everything else. You can only form the minds of reasoning animals upon Facts: nothing else will ever be of any service to them.

Gradgrind the headmaster in *Hard Times* by Charles Dickens

The most obvious aspect of any curriculum is the intended content to be covered. The content can form the focus of many conversations and public critique. In mathematics there are many discussions about content. Do we expect people to learn times tables up to 10 or 12 (or beyond)? Should we include algebra? What sort of geometry will be expected?

One key development within education was the project to categorise learning into various forms. Usually known as *Bloom's taxonomy of learning domains*, this divided educational activity into three domains:

> ❭ B1 *Cognitive*: concerned with mental skills and procedures

> ❭ B2 *Affective*: concerned with attitudes and emotions

> ❭ B3 *Psychomotor*: concerned with physical skills

Bloom's committee was concerned with dividing up learning domains into various levels to be able to describe higher education. Under the cognitive domain the first level consists of knowledge – these are the facts of Gradgrind – and moves through comprehension and application to higher-level skills such as evaluation.

In mathematics we may expect certain *facts* to be remembered, for example what the numeral symbols (1, 2, 3, 4 …) represent, what operations of small numbers produce (2 + 4, 3 × 6) and so on. We also expect learners to demonstrate comprehension and application. For example, in examinations (such as GCSE) assessors are looking for *comprehension* and allocate *method marks* for attempts to apply techniques whether the answers are right or wrong. In some cases, questions expect the selection and use of mathematical techniques to solve problems.

 ## TASK 3.1

A great deal of the content of mathematics would be seen as existing under the heading of B1 Cognitive. What aspects of mathematics might fall under domains B2 Affective and B3 Psychomotor?

Suggested responses to tasks can be found at the end of the book.

BEHAVIOURIST THEORIES

Theorists from the *behaviourist* school noted that organisms respond to stimuli in the environment and that behaviour can be studied in order to assess whether change has taken place. Skinner was one such theorist who conducted experiments with animals and believed that they could be trained to respond in different ways. He believed that it was possible to reinforce learning through the use of rewards or punishment. This perspective was expanded to include humans, with behaviourists treating people as 'black boxes' that should be observed for changes in behaviour. In essence, behaviourism is an attempt to use a scientific method to study the behaviour of learners without worrying too much about to the internal processes that produce such behaviours (the 'black box' is the opaque brain).

MATHEMATICS AS OBJECTIVES

Such behaviourist ideas led to teaching being focused around the intended outcomes of learning – in order to decide whether a particular series of activities had been successful. The notion of an *objectives model* of education was developed in the 1940s by educationalists such as Ralph Tyler (this is also called the *product model* by others). They were concerned with what students were able to demonstrate at the end of a programme of study. Teachers set learning objectives that can be measured and assess learning against a set of criteria. Products of learning such as homework or tests are marked and learners are rewarded for positive behavioural changes in different ways, such as praise or through certification.

Take multiplication as an example. There are a number of different objectives that a teacher might set for a class. Probably the most obvious objective is for learners to demonstrate that they are able to calculate the product of two numbers (usually with the level determining what sorts of numbers are involved, such as 2×4, 27×35 or 35.4×4.2). An alternative objective might be to identify situations in which multiplication is to be used. For example, consider an event in which there are six tables, each of which can seat 12 people. How many people can be seated?

 TASK 3.2

Take the mathematical concept of a fraction. When planning sessions, what objectives might you expect the learners to demonstrate?

Suggested responses to tasks can be found at the end of the book.

COGNITIVIST THEORIES

Theorists from the *cognitivist* school focus on the thinking and adjusting that the learner does when linking existing knowledge to new knowledge, and how that new material is organised and becomes part of the learner's understanding. The emphasis is on the learning process – unlike in behaviourism – along with the product of learning. The notion of a mental or cognitive map is used to describe how ideas fit together. For many educationalists a learner's attitude and feelings will also have a bearing on the learning process. Jean Piaget formulated his theory of cognitive development and described two ways that new knowledge is taken on board by learners. In the first way, a learner is introduced to new information which does not contradict existing ideas – in such circumstances learners are said to *assimilate* new knowledge by adding to current mental maps. In the second way, when new ideas contradict existing notions learners are required to *accommodate* this new knowledge by reorganising mental structures. Cognitivist ideas influenced discovery learning and investigations, of which more will be outlined later.

MATHEMATICS AS PROCESS

As educational thinking developed, and the focus on behavioural studies moved to more cognitive ideas, so the focus on curriculum moved from 'what' should be studied to 'how' a subject is undertaken (the *process model*). The introduction of GCSE Mathematics during the 1980s brought in 'using and applying' as a strand within mathematics and investigations as an activity type. One aspect of investigations involved learners outlining their thinking about how they undertake the investigation. More recently, Key and Functional skills introduced the three process skills of 'representing', 'analysing' and 'interpreting'. These process elements of mathematics have been developed and assessed in a number of different ways.

For example, consider the following three tasks.

Case 1. Which of the following calculations works out 30% of £300?

    (a) $30 \times 100 \div 300$

    (b) $100 \times 300 \div 30$

    (c) $30 \times 300 \div 100$

    (d) $100 \times 300 \times 30$

Case 2. Put the following calculations in order to find the sale price of a £400 television reduced by 25%.

    A $400 \div 4 = 100$

    B $25\% = \frac{1}{4}$

    C £400 – £100 = £300

    D To find 1/4 of an amount, divide by 4

Case 3. Toni wants to work out the 20% VAT on a plumbing job costing £150. Toni calculates the following:

    $150 \div 20 = 7.5$

    The VAT is £7.50

    Is Toni correct? Explain your answer.

A traditional, transmission-style approach to mathematics (see the discussion in the introduction) would usually involve asking learners to work out the calculations in each of these cases. This approach focuses on the calculations required and the answers produced. It has been noted that this often does not result in learners writing down their methods, even when instructed to do so – usually by requests to 'show your working'.

The three cases above take the focus away from answers and put the solution process to the fore. Case 1 looks at the numbers involved and potential operations between them, including some that individuals might think are true. It is possible to not have to calculate anything to identify an answer. Or one could work out each of the four options and make a decision from the results. Case 2 focuses on the order in which one undertakes calculations. This does not require any calculation to actually be carried out and there is no point in doing so as all the component parts are correct. Case 3 looks at the solution of another individual and invites the learner to take the role of a marker. Again an answer to the calculation is not sought but instead a comparison of what is seen with a correct response, with an explanation required at the end.

Despite these positive aspects of the tasks, there are some potential difficulties. Cases 1 and 2 expect the learners to think in the way of the writer. It is well known that learners often find it difficult to see even minor alternative calculations as equivalent. For example, in Case 1 a learner might think to calculate $300 \times 30 \div 100$, or want to write down a version in fraction form, and not see these as equivalent to (c). Also, in Case 2 some stages might be seen as irrelevant (for example (b)) or some might need expansion.

Case 3 requires an explanation and learners may not be sure of what counts as such an explanation.

This is not to suggest that these tasks are inappropriate – just that all tasks have advantages and disadvantages. There certainly can be difficulties in using these as items in assessments – Case 1 was often used in the multiple-choice assessments for Key Skills, whereas the approach in Case 3 is used within some functional mathematics assessments.

HUMANIST THEORIES

Theorists from the *humanist* school believe that learning should be directed by the learner and their needs. Learning is often seen as being for personal development, rather than for accreditation or financial benefit. Maslow postulated that people have a *hierarchy of needs* that must be met in a particular order and once this was achieved, an individual would be capable of achieving their potential. Within adult education, Malcolm Knowles developed the notion of *androgogy*, which suggested that adults learned differently to children and that motivations and prior knowledge are key to developing understanding. In both of these examples we see that non-mathematical elements play an important role in learning.

MATHEMATICS AS A PERSONAL DEVELOPMENT

There are a range of motivations for studying mathematics. Some – a minority of – learners study mathematics in order to help develop it as an academic subject when they progress to higher education. For others, mathematics can be studied in order to support another subject or area of study. In this section, we will consider the motivations for studying mathematics and how it can act as an element of personal development. It is known that learners often study mathematics in order to overcome a previous failure. This is not necessarily linked to other study. Some may even feel that it helps them with their everyday lives, for example in dealing with personal finance or supporting activities such as shopping or DIY. Others may want to help their children with their school work. There are those who wish to progress in their careers and there are those who want develop as active citizens. Maslow talked about individuals reaching a stage of self-actualisation.

TASK 3.3

What differences would there be in planning for individuals with the following motivations?

To overcome past failure

To help with everyday life

To help children with their school work

To help develop a career

To become active citizens

Suggested responses to tasks can be found at the end of the book.

A number of approaches to teaching have been developed that take into account learners' backgrounds, experiences and feelings. In mathematics education, there are the proposed multicultural and ethnomathematical approaches and ideas around critical mathematics education which will be discussed later in this chapter.

CONSTRUCTIVIST THEORIES

Before looking at some approaches to teaching it is helpful to consider the subcategory of learning theory labelled *constructivism* which has influenced the investigative and collaborative approaches outlined in the following sections.

Cognitive constructivism

According to *cognitive constructivism* theories, knowledge is actively constructed and is affected by the prior knowledge and experiences of the learner. Such constructions of knowledge are personal and subjective. Jerome Bruner was an exponent who believed that learners should be provided with opportunities to explore and discover truths and therefore formulate their own theories with reference to their pre-existing knowledge. He believed that in this way learners would come to a deeper understanding of concepts and would remember what they had learned. A teacher is seen as facilitating learner discussions in which each participant generates their own knowledge from learner interactions. Assessment is seen as part of the learning process and learners are involved through activities involving peer and self-assessment. The focus is on the process of learning rather than products such as formal accreditation.

Along with cognitive contructivism came the idea of *discovery learning* in which learners are encouraged to explore ideas with little guidance. Discovery learning involves activities, investigations or experiments in which learners are given a task or problem to be solved. Discovery learning can be, and often is, carried out in small groups in the classroom and learners can support each other but the focus is how *individuals* make sense of the learning experiences.

There are notions of a more *guided discovery* in which more structured support is provided by a teacher. A series of questions or a task sheet may be used to help guide the learner towards where the teacher wants them to get to without actually telling them the answers.

Social constructivism

Social constructivism describes how learning grows out of social encounters. Lev Vygotsky found that learning was better achieved through working with a *more knowledgeable other* (MKO). Engagement with the MKO was found to help learners to construct their own understanding. Vygotsky postulated the *Zone of Proximal Development* (ZPD) which represents a learner's potential that can be reached with the support of an MKO, this zone being just beyond their current attainment level. He believed that language is a vital part of the learning process as it is the tool used to develop and articulate ideas.

More broadly, social constructivist theories deal with how individuals learn when they collaborate with others. It is believed that interacting with others allows individuals to engage and analyse the ideas of others and to articulate and reflect on their own ideas. This, in turn, is felt to help individuals within the group to build a deeper understanding of the learning material.

APPROACHES TO TEACHING MATHEMATICS/NUMERACY

Investigative approaches

Investigative approaches involve learners working in small groups with a given mathematical problem to resolve. These approaches developed alongside the theoretical notion of constructivism within education theory and are associated with discovery learning.

As was noted in Chapter 2, investigations were introduced into mathematics education to allow learners to experience the mathematical process in a similar way to an academic mathematician. Until fairly recently, investigations formed part of the GCSE Mathematics assessment within the assessed element of coursework for all examination boards. In this case, the work was designed to be carried out individually on topics such as number patterns, statistics and probability. Despite this well-known link to GCSE accreditation, investigations were originally developed and employed as a learning approach not necessarily related to assessment. It is investigation as a teaching approach, however, that we will now focus on.

An investigation may be introduced by asking a simple question.

> *Example*: Do rectangles with longer perimeters have larger areas than those with shorter perimeters?

Investigations are developed to take learners through the process of mathematical problem solving. This will involve learners planning what to do, trialling different approaches, reflecting on their results, coming to conclusions and presenting their findings. Using such a full cycle of learning is felt to be valuable as learners are believed to reach a fundamental understanding of the topic being investigated. This deeper understanding is thought to be longer lasting, therefore providing a better learning experience. Another reason that mathematical investigations may have long-term gains is that they often result in the learners making connections with mathematical areas other than the primary focus of the investigation as they explore and search for solutions. This will help to develop skills that can be transferred to other areas of mathematics and even beyond the subject. For example, students may develop the higher-order skills of reasoning, analysis and problem solving. In presenting their findings and working with others they should also develop their written and oral communication skills, in particular extending their active mathematical vocabulary. They may also develop their ICT skills when working on the investigation.

Teachers who use investigation as a teaching approach need to consider how to prepare and support learners to get the most out of the investigation, guiding them towards a solution if necessary while ensuring that the learners are the ones doing the discovery. Guidance may take the form of carefully worded questions that could help the learner who doesn't know where to begin to start looking for a solution in appropriate places.

Returning to the example:

> Do rectangles with longer perimeters have larger areas than those with shorter perimeters?

With this investigation the teacher may wish to check that the learners are clear about the meanings of certain mathematical terms such as rectangle, area and perimeter. On the other hand, if learners are working in groups, discussion around these terms and how to calculate area and perimeter may provide additional learning opportunities. The teacher may also decide to provide material such as squared paper to support the task.

Some other issues to consider:

> ❱ It is advisable for the teacher to carry out the investigation themselves before asking the learners to do it, so that they are aware of some potential pitfalls and other issues such as resources required (including time). Be careful, though; teachers should also be open to learners using different strategies, or possibly solutions, to themselves.

> ❱ Learners will need to have access to the required tools, including physical materials, as well as having the basic skills and knowledge necessary to conduct the investigation.

> ❱ If learners are set an investigation to carry out in groups, the teacher should check that all group members are contributing towards the process and are therefore all benefitting from it.

> ❱ Investigations can be open-ended with a number of solutions, which allows learners to access them at different levels. However, learners should be very clear about what is expected of them, any deadlines and how their performance will be judged if appropriate.

> ❱ The language used to present a task will need to be checked for appropriateness for the particular group of learners. For example, an instruction to 'explore the relationship between circle diameter and circumference' may not be accessible to some learners who may require simpler language and possibly a more guided approach. In this case, the

teacher could provide some circular objects of different sizes and other materials such as paper, string, tape measures and rulers. The teacher could then suggest that the learners find a way of measuring across and around the objects, comparing the two measurements for different objects and seeing what they notice.

> There are open-ended investigations that do not have fixed responses. These can make for a less threatening type of investigation as long as the learners realise you are not expecting a correct answer. It also opens up the variety of approaches and skills that learners will employ to try to reach a reasonable solution. For example, the Fermi-style problem (cf Chapter 2) 'How many people could fit into this room?' may lead to discussions of average size of person, dimensions of the room, area and capacity. Different learners may make different assumptions, such as all the people are adults or they must all stand on the floor, which makes for interesting discussions during the feedback stage.

> Higher-level students can be given investigations which allow them to use higher-level skills. For example, a task to 'find the best estimate for the height of this room' will probably result in a variety of strategies being used to find a good estimate and, if posed after a lesson on trigonometry, should lead some to use it to estimate the height. This could progress to a discussion about the application of trigonometry in real-life situations.

> Teachers should allow sufficient time for learners to present their findings and for class discussion of results.

Before moving from investigations there are two potential concerns that we would like to raise. Adult learners may have negative experiences of investigations from school or may have never carried out an investigation, so it is important to try to make the process engaging and to explain how they can go about the investigation without 'giving the game away'. Open-ended problems may be frightening for many learners who are used to a more structured, teacher-led approach. Also bear in mind that an investigation may be attractive to learners who prefer more creative subjects and who do not enjoy the more traditional, transmission-style approach where they are expected to arrive at a particular 'correct answer'. It may be a good idea to start with something simple that could be completed in a short space of time if this is the first investigation for the group. You could also place learners in groups, ensuring that each group has a member who has completed an investigation before.

With this type of discovery learning, there is a concern that students may not recognise some of the misconceptions that arise during their investigations. Groups may think they have discovered a rule that works and latch onto it without testing it fully. For example, some learners when carrying out the example perimeter investigation may simply compare two rectangles and want to come to a conclusion on the basis of that single comparison. In that case, the teacher would probably suggest trying some other rectangles to test out their conclusion and may make a note to discuss the need to test out any early findings in different cases, before being satisfied that they have discovered a general rule.

Another simple investigation is to have a set of shapes and pose the following question: *'In how many ways can these shapes be divided into two equal halves?'* This investigation could be differentiated by giving more complex shapes to learners who are working at a higher level. The task could also be made more difficult by just naming shapes to be considered, or easier by providing images of the shapes together with pairs of scissors.

Collaborative approaches

Collaborative approaches links closely to social constructivist learning theories noted earlier. Such perspectives suggest that learners will learn more when collaborating in groups, through discussion, than they would when working individually.

Example

One type of activity is the 'odd one out' task. The idea of the task is not to identify a single 'odd one out' but rather to note how each case could be seen as the odd one out.

Look at the following cards (with the numbers 2, 3 and 6) and decide how each of these could be the 'odd one out'.

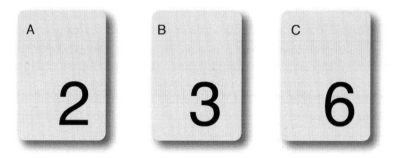

In this example, A could be the odd one out because the other two are multiples of 3, B could be the odd one out because the other two are even, or C could be the odd one out because the other two are prime numbers.

By having lots of examples like this, learners would develop their notions of types of number through discussion.

Malcolm Swan talks about the need for collaborative activities to be rich tasks, meaning that they should be designed to throw up misconceptions and to create *cognitive conflict* which through discussion will be resolved. Cognitive conflict occurs when someone thinks that they understand a notion but is presented with contradictory information. The resolution to this conflict would require the process of *accommodation* outlined by Piaget (see earlier). For example, it is not uncommon for learners to believe that the more digits a number has, the greater its value. If they are presented with a task that highlights this misconception and demonstrates that it does not always hold true, they will initially struggle with what they are experiencing. Resolution of this conflict is believed to result in a fundamental understanding of the concept replacing the previously held misconception.

Collaborative tasks should be selected or designed and managed to promote discussion of the related mathematical ideas. In addition to choosing a task that will do this, some language preparation may be necessary so that the individual learners actively participate in the discussions. For example, if a teacher plans for the learners to work in small groups on a collaborative task relating to shape, they may preface it with a whole-group activity that includes a focus on the language of shape. When learners are truly collaborating on a task they will naturally discuss what they are doing using mathematical register. Register is the set of words and phrases that are associated with a particular subject or context, together with patterns of speech that are particular to that subject or context. These discussions will help to expand their vocabulary and ability to talk mathematically, in addition to sharing and comparing ideas about the mathematical concepts themselves. The belief is that language is the key to understanding because thinking is language-based. Articulation of ideas therefore promotes understanding.

Card-based activities are used for sorting information into different categories or matching pairs or groups of cards together according to their properties. They may also involve ordering values, finding the odd one out of a group of images, numerical or text-based information, or identifying if statements are true or false. They can be used to present a significant number of mathematical ideas in a simple way that is more accessible than text-heavy worksheets. However, they can be designed to deliberately provoke discussion about concepts that are known to be problematic, or to draw attention to particular issues. The following example focuses on some key properties of numbers.

Example

Sort the following into statements that are TRUE and those that are FALSE.

Adding two odd numbers gives an odd number	Adding two even numbers gives an even number	Adding an even number and an odd number makes an even number
Subtracting two even numbers makes an odd number	Subtracting two odd numbers makes an even number	Subtracting an even number from an odd number makes an odd number
Multiplying two even numbers makes an even number	Multiplying an odd and an even number makes an odd number	Multiplying two odd numbers makes an even number

You could choose to change the categories to 'always true', 'sometimes true' and 'never true'. This might help if learners are not happy with using negative numbers that may be the result of subtractions.

Learners tend to enjoy such card activities and find them less threatening than written exercises. This is partly because the activities lend themselves to collaborative working and partly because their suggested solutions can appear less permanent than a written answer. Learners can test out placement of a card and discuss whether they agree that it is correct. If they don't agree, they can easily shift the card to a new position and discuss it again. During this process there should be a lot of discussion and sharing of ideas. These activities usually involve the checking of solutions – either by comparing a group's solution with other groups which leads to further discussion, or by checking against an answer sheet. Comparing responses is the responsibility of the group rather than individuals and it should lead to further discussion of why the mistakes were made. Finally, the teacher will take feedback on the activity from each group where any issues or misconceptions can be discussed.

Games and puzzle activities provide potential candidates for collaborative tasks. Learners tend to regard games as something different to learning mathematics, but they can be based on challenging mathematical concepts. Learners often find playing games fun as they feel they are not being tested and they can relax and enjoy playing the game with others. This, in turn, can promote an easier flow of conversation between the learners, but if the game requires the use of mathematical skills and knowledge they will still be using mathematical language. Games also promote active collaboration between group members because the situation feels less threatening than, for example, working on a paper-based task together.

> *Example*
>
> You may have played a memory game where cards are placed on a table with the text hidden. Participants then choose to turn over two cards at a time. If the cards match in some way then the participant keeps the cards and tries again. If the cards don't match they are turned back over and the play moves to the next person. The winner is the person with most cards. The cards could have equivalencies on them, such as fractions and percentages.

Collaborative learning isn't as simple as getting learners to work in small groups on a task or activity. This may in itself result in some learning outcomes for some of the groups but there are potential barriers to collaborative learning resulting from simply asking learners to work in pairs or threes. One such barrier is that learners may ignore the instruction to work together or talk to each other about it, with the result that one learner completes the task while others look on or they all do their own thing with very little conversation. For collaborative learning to be effective, there must be discussion and exchange of ideas about the task and the underlying mathematical concepts involved.

> ❯ Collaborative learning can be assisted by having only one learning resource per group and ensuring that the group size is small enough to encourage active participation by all group members.

> ❯ Grouping may need careful management – there may be some learners who do not work well together, for example. The teacher may also avoid placing very quiet learners with dominant ones or may suggest that each group member has a different role within the group.

〉 A strategy that the teacher could use to encourage active participation of all group members is to inform them that each member will be required to provide some feedback on the task.

Collaborative strategies may need to be gently introduced to those in a class who seem resistant to the idea or who have never learned mathematics in that way before – possibly due to different cultural or educational backgrounds. It will probably take time for those in a group new to the ideas of collaborative learning to understand how to work in this way.

Another possible barrier or concern is that some learners may mislead other less confident ones, therefore confusing them. This should not constitute a problem as long as the teacher takes feedback on the task from each group where any errors will come to light and can be addressed.

Intervention may take the form of questioning group members about how the cards have been placed, or suggesting strategies such as taking turns to place the cards and explaining to each other why they are placing them in a certain way. If some group members feel less confident about placing the cards they should be encouraged to ask questions rather than passively observing.

Multicultural and ethnomathematical approaches

Multicultural and *ethnomathematical* approaches take on board constructivist and humanist ideas. Notions of *ethnomathematics* are somewhat varied but all involve the exploration of some form of cultural practice from a mathematical perspective. The intent of such approaches is to value, use and share knowledge that is familiar and relevant to individuals and bridge to more formal mathematics. We use the term *multicultural mathematics* in a more general sense to identify mathematics used in various cultures, whether or not it is familiar to learners. Such approaches take into account the context within which learning takes place, as well as considering cognitive aspects of learning.

Example

This activity comes from Paulos Gerdes and his work on ethnomathematics.

In Mozambique, square mats are turned into baskets by weaving strips.

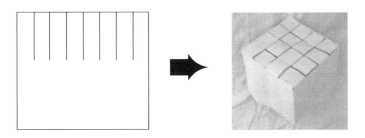

Make your own using square card, cutting into 8 equal strips as shown above. What shape is made when you weave the strips?

Such activities are intended to be relevant to the range of cultural backgrounds represented within a group of learners. This requires the teacher to have an awareness of, and respect for, the different cultural backgrounds of their group of learners, as well as an awareness of how their own culture has affected their views of mathematics as both a learner and a teacher.

Gelsa Knijnik has described how groups of peasant workers in Brazil develop their notions of rounding in relation to what they do in real life. There is no abstract rule, such as 'If the following digit is greater than 5 then round up, else round down'. For these workers, you round up when you want to make sure you have enough money for an exchange, or round down when considering income. Such mathematics is context bound and to talk of general rules would be inappropriate.

In using ethnomathematical approaches:

❱ the teacher will need to be open to the idea that problems can be solved in different ways, using different strategies and methods, rather than there being one correct method that should be applied in all cases. They should also develop their knowledge of a range of alternative strategies – this should be regarded as an essential part of their training as a teacher of mathematics in a multicultural society.

❱ learners, too, may need convincing that the methods they use which they may regard as informal or not 'real mathematics' (possibly because they have previously been told so) are just as valid as what they see as the correct methods. This approach not only supports development of mathematical skills and knowledge of cultural minorities but it also enables all learners to build a 'toolbox' of different strategies which supports a flexible approach to problem solving and helps to develop conceptual understanding as the various methods are discussed and compared.

Mathematical ability becomes less reliant on remembering and recalling a set of rules and more about finding and using a strategy that works for them and suits the problem.

One multicultural approach is the use of the history of mathematics. Despite some texts which emphasise Western contributions to mathematics, the history of mathematics is truly a multicultural phenomenon. It also adds interest to the subject and helps learners to see why certain mathematical practices arose. For example, when teaching place value the teacher may present information about different number systems that arose in ancient civilisations such as the Babylonians, Egyptians and Romans. Comparison of systems that are based around having place value with those that do not can prove to be interesting and useful in terms of explaining the concept of place value in our decimal number system.

Mathematical topics require an understanding of a set of terms and phrases that may be unfamiliar to learners. Looking at the roots of the words and discussing how current meanings have come about can help learners to remember the meanings of terms by highlighting word patterns. Indeed, it is known that language can pose a barrier to native English speakers as well as to learners whose first language is not English. For example, it is considered good practice to ask learners who speak other languages to share their words for mathematical terms as it values their languages and can illustrate different patterns, for example how numbers one to twenty are written in different languages. The topic of shape has words for shape names and for describing properties that may look difficult to decipher. Drawing on the knowledge that learners bring from their own lives and experience and including discussions can allow attention to be focused on meaning. For example, identifying that 'poly' means many and 'gon' is derived from the Greek word for angle may prove helpful. When teaching measurement, it can be helpful to focus on the prefixes of the terms in the metric system. Discussing different systems and how different languages use terminology can bring to the fore the links between examples of words that contain those prefixes, such as century meaning a hundred years.

Another multicultural approach is to avoid only using problems that are culturally specific to the UK. Other cultural groups use different cultural practices that can be represented in mathematical problems, in addition to using names and images taken from other cultures. This helps to make the problems more familiar and relevant to different groups of learners and can help prepare them for assessments that use context-based written problems. Resources linked to other cultures can also be used in support learning; for example, the abacus has strong links to Chinese culture although it was used by other civilisations such as the Babylonians. It can be used to demonstrate counting and calculation strategies.

Critical mathematical approaches

Critical mathematical approaches generally takes the form of discussion using information drawn from the media or other sources. The teacher tries to ensure that the context of such problems is realistic and relevant to the learners. This is different to the tokenistic contexts that have historically been used in classroom mathematics problems.

As with the ethnomathematical approaches, these approaches are linked to constructivist and humanist theoretical perspectives as they focus on finding out what different learners know already about a real context and assessing what they might need to know. Teachers who adopt this type of approach tend to make overt links between what they are teaching and its uses. For some it is believed to be motivational and to support application of the skills. For others, this is a political mission to encourage active citizens and mathematics is a secondary issue. Learners are encouraged to question and be critical of what they observe or are told, rather than passively accepting information as the truth. This is achieved by boosting learner confidence and presenting different viewpoints as well as highlighting the fallibility of examples of mathematical information. An example of this is demonstrating how statistics can be used to distort the message being transmitted.

Many advocates of such approaches do so in order to produce politically aware citizens. The idea of taking some very real information, discussing what the information might mean and being critical about the way it is discussed in the sources, is the point of education for many. Having said this, there are implications. To what extent would all learners feel comfortable with such an approach – or all teachers?

Consider Task 3.4. What issues may be raised by this activity?

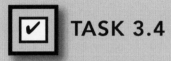

TASK 3.4

Look at the following data about births in London. What do you notice?

Area name	2009		2012	
	Live births	GFR[1]	Live births	GFR[1]
Barking and Dagenham	3,624	90.5	3,957	90.0
Barnet	5,286	70.7	5,585	68.3
Bexley	3,029	65.1	3,076	64.9
Brent	5,132	89.7	5,340	72.1
Bromley	4,104	64.9	4,140	66.5
Camden	3,094	47.3	2,944	48.2
Croydon	5,235	69.7	5,884	73.0
Ealing	5,638	77.1	5,847	73.7
Enfield	4,828	76.1	5,094	71.9
Greenwich	4,480	82.7	4,624	74.6
Hackney and City of London	4,574	76.0	4,585	63.7
Hammersmith and Fulham	2,841	62.3	2,646	52.5
Haringey	4,193	74.2	4,209	64.7
Harrow	3,265	66.9	3,585	69.6
Havering	2,697	59.1	2,888	61.9
Hillingdon	4,207	71.1	4,536	72.5
Hounslow	4,297	77.9	4,621	76.2
Islington	2,983	55.1	2,988	49.5
Kensington and Chelsea	2,227	54.6	2,024	53.9
Kingston upon Thames	2,321	58.8	2,328	60.6
Lambeth	4,863	64.2	4,825	56.0
Lewisham	4,888	73.8	5,095	72.1
Merton	3,462	70.8	3,476	72.6
Newham	6,003	103.0	6,426	82.3
Redbridge	4,253	71.9	4,792	75.5
Richmond upon Thames	2,859	68.3	2,916	72.1
Southwark	4,873	64.8	5,030	62.7
Sutton	2,786	67.2	2,708	66.0
Tower Hamlets	4,337	66.1	4,784	62.1
Waltham Forest	4,533	87.3	4,832	77.5
Wandsworth	5,335	62.2	5,451	59.8
Westminster	2,998	42.6	2,950	51.6
Inner London	53,209	64.9	53,957	60.8

	2009		2012	
Outer London	76,036	73.2	80,229	71.8
North East	29,776	58.1	30,291	60.6
North West	87,549	63.4	89,211	64.1
Yorkshire and the Humber	66,358	61.9	67,408	64.2
East Midlands	53,746	60.6	55,645	63.1
West Midlands	71,042	65.9	73,940	67.2
East	71,335	63.7	74,571	66.5
London	129,245	69.5	134,186	67.0
South East	103,669	62.5	107,858	64.5
South West	58,338	59.9	61,131	63.1
Wales	34,937	60.9	35,238	61.2
England	671,058	63.7	694,241	64.9
England and Wales	706,248	63.6	729,674	64.8

1. The General Fertility Rate (GFR) is the number of live births per 1,000 women aged 15–44.
Suggested responses to tasks can be found at the end of the book.

This type of approach deals with the importance of mathematics in equipping adults to take a full and active role in society. Links are made to everyday mathematics and the teacher aims to develop skills and knowledge that produce critical citizens.

But there is a potential for a range of difficult discussions around sensitive topics. This might be very real to learners in the groups but may also encourage inappropriate contributions. For some involved in critical mathematics education, the point would be to address these issues head on. In encouraging your learners to use mathematics to question information and views presented you are in danger of imposing your own agenda. In any case, such discussions raise delicate matters and should be handled sensitively.

One final point to consider here is the relationship to national awards. The current national mathematics qualifications for adults are based on notions of *functional mathematics*; this curriculum came about because of a perceived need for a curriculum more related to the real world. Nevertheless, such schemes are unlikely to deal with such sensitive matters that others feel are important to discuss.

Another example

Consider the following extract from a BBC News article. What points are made in the article? What do you think? How does the data help your thinking?

Alcohol consumption 'continues to fall'

Alcohol consumption in 2009 saw the sharpest year-on-year decline since 1948, figures from the British Beer and Pub Association (BBPA) suggest.

The data showed a 6% fall in 2009 – the fourth annual drop in five years.

The association said UK drinkers were now consuming 13% less alcohol than in 2004, below the EU average.

Pubs, bars, off-licences, restaurants and supermarkets all saw alcohol sales fall, the HM Revenue and Customs data from UK producers and importers showed.

It is thought the decline may be due to the effect of the recession on spending, but could also be a sign that messages about responsible drinking have affected drinking habits.

The organisation said UK taxes on beer remained the second highest duty rate in EU – 10 times higher than in Germany and seven times higher than in France.

2008 alcohol consumption in Europe (litres per head)

Czech Republic – 12.3

Austria – 10.4

Lithuania – 10.1

Germany – 10.0

Spain/Hungary – 9.8

Portugal/Slovakia/Denmark – 9.3

Poland – 9.8

Belgium/Luxembourg – 8.5

UK – 8.4

Finland/Greece – 7.6

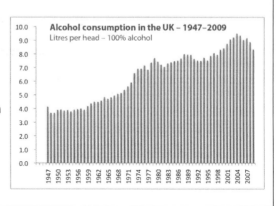

Alcohol consumption in the UK – 1947–2009
Litres per head – 100% alcohol

Text © BBC News (www.bbc.co.uk/news/uk-11170814); data © BBPA, from *BBPA Statistical Handbook 2010*.

SUMMARY

This chapter has outlined some key theoretical perspectives and related these to some practical aspects of teaching mathematics/numeracy. In addition, we have identified four categories of approaches to learning and related these to theoretical perspectives. Investigative approaches are intended to focus on mathematical problem solving, whereas collaborative approaches focus on how discussion aids concept development. Multicultural and ethnomathematical approaches value the knowledge that learners bring from their backgrounds and attempts to link this knowledge to mathematical ideas. Critical mathematics education moves the focus away from the subject itself to encouraging individuals to use mathematical ideas in becoming an active citizen. It's not that you will necessarily choose one of these positions but you may be mostly influenced by one or the other. It may also be the case that different parts of a course you deliver may be more consistent with different approaches.

FURTHER READING

Coben, D., Brown, M., Rhodes, V., Swain, J., Ananiadou, K. and Brown, P. (2007) *Effective Teaching and Learning: Numeracy*. London: NRDC.

Griffiths G. and Stone, R. (eds) (2013) *Teaching Adult Numeracy: Principles and Practice*. Maidenhead: Open University Press.

Swan, M. (2006) *Collaborative Learning in Mathematics: A Challenge to Our Beliefs and Practices*. Leicester: NIACE.

4 The teaching and learning cycle

INTRODUCTION

(Cartoon reproduced from *Improving Learning in Mathematics: Challenges and Strategies* [DfES, 2005] under the Open Government Licence v 3.0)

This chapter builds on some of the information in previous chapters. It covers the range of activities that effective teaching depends upon, such as the different styles of assessment used in lessons, planning learning, formative and summative feedback and checking learning.

Once a teacher is established in the classroom, one of their main concerns will be to ensure that learning has taken place. There is a well-known cartoon showing a boy with his dog (above). The boy says he has taught his dog to whistle. When his friend points out the dog is not whistling, the boy defends himself by saying, 'I said I had *taught* my dog to whistle, I didn't say he'd *learned*!' Teaching can often be like that, with learners apparently refusing to learn however well you teach them.

We use the phrase 'teaching and learning cycle' to describe the complex interweaving of planning, teaching and assessment which, in turn, leads to learning. There are many different versions of the teaching and learning cycle, either more or less complex and favouring one or other of the theoretical approaches outlined in Chapter 3. However, all follow the basic form of: initial assessment, lesson planning, lesson teaching, formative assessment and summative assessment. They may have four, five or six stages, but all illustrate this same interplay of: assessments, planning and teaching.

AN EXAMPLE OF A TEACHING AND LEARNING CYCLE

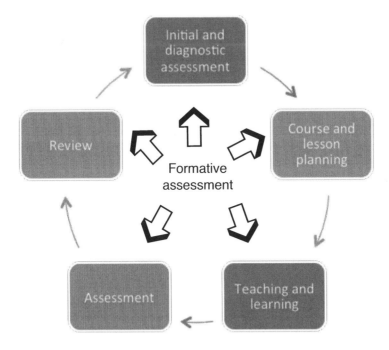

One of the most important things to understand when teaching post-16 learners is that they do not come to a class without any knowledge of mathematics. All have done some mathematics at school, many have prospered in adult life, have been in work or are raising families. They may state that they know no mathematics, but they do. The challenge for the teacher is find out what they do and don't know, what skills and understanding they have and what misconceptions may be holding them back.

Black and Wiliam established that future learning could be formed by assessment. They suggested that assessment should primarily be carried out to inform and direct learning and that assessment should be *for* learning rather than simply *of* learning. This has had a profound impact on teaching practice. It means that as they teach, teachers are expected to be continuously assessing, but may also utilise specific opportunities to bring assessment to the fore. Assessment can be both implicit and explicit.

Certain assessment processes, in particular self-assessment and peer assessment, contribute to building learners' own sense of autonomy and their ability to take responsibility for their own learning and actions. Increasingly helping learners become autonomous is being seen as an essential element of successful teaching.

INITIAL AND DIAGNOSTIC ASSESSMENT

In working for most colleges or training providers you will be expected to conduct an initial and/or diagnostic assessment. The purpose of these assessments is to build up a picture of the learner's mathematics needs, abilities and goals. Ideally this process will include a face-to-face element, allowing the teacher to understand what the learner is aiming to achieve and providing an opportunity to explain to the learner what the teacher expects of them.

Initial assessment should identify:

❯ aims of the learner, including short- and long-term goals

❯ previous educational background

❯ preferred ways of learning

❯ any language preferences

❯ any necessary additional support needs

The emphasis should be on what the learner can do and wants to do, not what they can't do or struggle with.

A diagnostic assessment provides a more in-depth profile of the learner's knowledge and abilities. These assessments are often computer-based systems, and may be quite long and stressful for the unconfident learner. Care needs to be taken that the act of taking a diagnostic assessment does not put them off entering learning at all!

TASK 4.1

Consider the advantages and disadvantages of the following as part of an initial or diagnostic process:

a) What is 20% of 250?

b) A plumber charges £150 for a job and needs to add VAT at 20% to the job. What would the VAT be?

c) Read the following question. Don't work out the answer but consider how confident you are in being able to answer the question.

A washing machine worth £500 is in a 20% sale, what is the discount?

Circle the statement which best describes your views:

Not confident at all, not very confident, a little confident, very confident.

d) Write down what you know about working out percentages of amounts.

Suggested responses to tasks can be found at the end of the book.

A worthwhile activity is to identify what initial and diagnostic assessments are used in the organisation you work for or have a placement with. Are the same assessments used for all levels of mathematics and numeracy? What happens to the scores from the assessments? Do you think they are administered in the best way?

Looking at the initial and diagnostic assessments allows the teacher to create an individual learning plan, or ILP. This document will identify the main areas where the learner needs to make progress during the course. The ILP defines a great deal of the content that the learner will be given, and so the diagnostic assessment can be viewed as 'assessment for learning' in that the assessment informs what areas should be taught. The use of ILPs has been challenged by some. This is partly due to the paperwork demands upon teachers but also because it can be difficult for learners to truly engage with the language of education. Nevertheless, most practitioners would agree that teachers should understand and assist the personal learning journeys of their learners.

The ILP is also the main process for allowing 'personalised learning'. It is often the case in the sector that teachers will be expected to negotiate a personalised learning plan with each learner, based on the diagnostic assessment but also including the learner's own motivations and ambitions. This again allows the learner to take ownership of their own learning, although as was noted above it may be difficult for some learners to engage with the language involved.

ELEMENTS THAT MAKE UP AN ILP

Initial/diagnostic assessments:

> Areas of strength and areas of skills deficit

> Additional support needs

Aims and targets:

> Long-term and short-term targets with dates

> Qualification aim or final outcome

Work related issues

Information from the learner:

> Long-term aims and ambitions

> Priorities for learning

> Preferred ways of learning

> Interests

The numeracy or mathematics ILP may not be the only ILP belonging to an individual learner. With so many individual pathways for those embarking on vocational education it is quite common for learners to have programme-wide ILPs covering their entire programme of study.

As previously noted, post-compulsory numeracy and mathematics classes come in all shapes and sizes. The above approach is something of an ideal and while it might work with, perhaps, a daytime class of mature adults, there are many practical problems with creating ILPs for all learners in a further education college group of 25 retaking their GCSE exams or needing Functional Mathematics as part of a vocational programme. In this case the best you might be able to manage is a 'group' ILP, creating a profile of the group as a whole in order to inform priorities for the scheme of work. Teaching in this sector is so often a matter of improvising and prioritising, managing the best way of approaching teaching in a wide range of contexts.

 TASK 4.2

Consider the following possible statements proposed as priorities for learning and comment on how appropriate they are for entries in an individual learning plan.

a. Work on multiplying

b. Be able to convert between metric lengths

c. Use bar charts

Suggested responses to tasks can be found at the end of the book.

FORMATIVE ASSESSMENT

However detailed the initial and diagnostic assessments, the teacher should always be looking to understand the extent of learners' knowledge on any topic; indeed, it could be said that the teacher of post-16 mathematics should be assessing learners' knowledge, in one way or another, at all times.

Research shows that teaching is more effective when it assesses and uses prior learning to adapt to the needs of learners. Prior learning may be uncovered through any activity that offers learners opportunities to express their understanding. It is generally recognised that assessment is most helpful when it is both ongoing and integral to teaching and learning.

The purpose of formative assessment is to provide a continuous process which charts achievement, identifies areas for development and indicates next steps for teachers and learners. It can be either formal or informal, or a combination.

The key elements that need to be in place to support formative assessment are:

❭ a classroom culture that encourages interaction and the use of assessment tools

❭ establishment of individual learning goals

❭ teaching that meets diverse student needs

❱ varied approaches to assessing student understanding

❱ effective feedback

❱ the involvement of students in the learning process

QUESTIONING

Effective questioning is a key technique for teaching. While closed questions have a role in mathematics teaching, teachers should avoid too many leading and rhetorical questions which may discourage unconfident learners and stifle debate and discussion.

Using open questions which allow learners to contribute as much as they are confident allows the teacher to understand how developed their learners' knowledge is. Open questions often start with 'Why...?', 'Give me an example...' or 'Can you justify that?'

Teachers need to allow plenty of time for answers, especially if the questions are open and challenging. Research suggests that teachers often leave only a second for learners to answer before answering the question themselves or firing off a follow-up question! Anyone who has filmed in a classroom knows that teachers are not very confident about leaving silences unfilled.

Open questioning is also a good way of uncovering misconceptions, which may be preventing some of the learners from understanding key concepts. These concepts need to be explored with the whole group. Mistakes are often the result of consistent, alternative interpretations of mathematical ideas. Typical misconceptions might be that division always makes a number smaller or that the longer the decimal expansion, the smaller the number. Discussion confronting the inconsistencies which arise causes cognitive conflict which can help to resolve those misconceptions.

 TASK 4.3

Read the following closed questions and write a related open question.

a. What is 12 times 6?

b. How many millimetres are in a metre?

c. How many sides does a pentagon have?

Suggested responses to tasks can be found at the end of the book.

DIALOGUE

Open discussion in the classroom, with learners treated as equals, allows learners to express their understanding, raise doubts and uncertainties or establish ownership of concepts. Dialogue helps to build rapport in the classroom, promotes participatory and democratic learning and so encourages learners to build their autonomy. The key element for promoting dialogue is mutual respect between the teacher and the learners.

DISCUSSION AND GROUP WORK

One of the advantages of small group work is that it allows the teacher to listen to conversations between learners as they explore concepts. Teachers should try and listen to what learners are saying and fight the impulse to 'tell them how it is'. By listening to learners working, the teacher can gain an understanding of the problems learners have and how widespread they may be. Listening to mathematically focused conversations is part of a powerful assessment process.

CREATING POSTERS

The teacher can assess groups through tasks such as creating posters. A poster on the topic of, say, 'What we know about circles' can provide a good indication of the knowledge of learners in the group and a quick photograph taken on your mobile phone acts as a record. Following discussion, peer teaching and further assessment tasks can provide evidence of 'distance travelled'.

Previously prepared presentations, or presentations based on group work done during the lesson also provide opportunities to assess the knowledge and understanding of both individuals and a whole group.

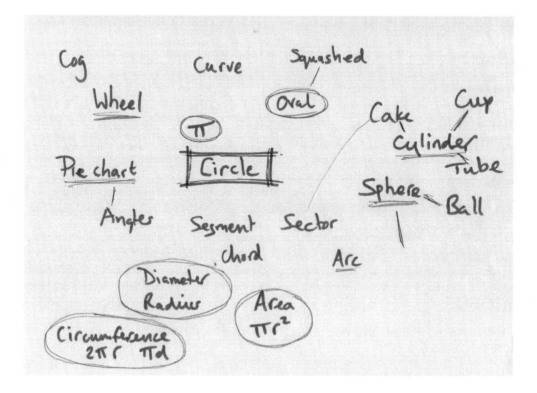

Perimeter & Area

Always

Slide the top corner of a triangle from left to right. The area of the triangle stays the same.

Drop vertical. Each triangle is half rectangle.

Draw a trapezium and draw its diagonals. The shape is now split into four triangles. Exactly two of these triangles are equal in area.

Same base
Same vertical height
→ same area

Sometimes

When you cut a piece off a shape you reduce its area and perimeter.

area smaller
perimeter bigger

area smaller
perimeter bigger

area smaller
perimeter smaller

When you cut a shape and rearrange the pieces, the area and perimeter stay the same.

| A | B | → | B | A |

Same

Never

If a square and a rectangle have the same perimeter, the square has the smaller area.

area = 36
per = 24

area = 35
per = 24

area = 32
per = 24

not the same perimeter

not the same

same

PEER AND SELF-ASSESSMENT

Inexperienced teachers often feel that the only meaningful assessment is done by themselves, and can wear themselves out with constant assessment exercises and the related record-keeping. However, both peer- and self-assessment processes are extremely valuable. Indeed, if the goal is to build autonomous learners, an understanding of self-assessment is crucial. Asking learners to set each other problems, devise ideal solutions and even devise marking schemes are valuable ways of assessing understanding of topics, enhancing learning and allowing learners to make connections between topics and between class work and examinations.

Using these methods of formative assessment regularly and consistently over time will make it possible to foster a collaborative culture in which learners take responsibility both for their own learning and that of their peers.

✓ TASK 4.4

Produce your own poster or presentation showing the advantages of using peer or self-assessments with your learners.

Suggested responses to tasks can be found at the end of the book.

INTERIM ASSESSMENTS

However expert the teacher may be at informal assessment, at some point a more formal assessment will need to be carried out, perhaps at the end of a scheme of work, to record progress for administrative purposes or as exam practice. While learners will undoubtedly see these as 'maths tests', teachers should try to make these assessments as interesting as possible and aim to avoid too many 'rows of questions with ticks or crosses'. Many learners who struggled at school will have a deeply negative response to seeing crosses against their work, while others easily become addicted to neat lines of ticks. But remember that you usually learn more about someone's thinking from the mistakes than from what they get right. If a learner has correctly answered three questions of the form:

$x + y = c$ What is x?

the chances are they can do thirty or even 300 of them. To understand their depth of knowledge you need to make the question more challenging:

$x + y = c - d$ What is x?

$2x + y = c$ What is x?

$3x + 2y = c - d$ What is x?

When the learner starts to make mistakes, the teacher will have an insight into their knowledge and understanding. It takes relatively few questions to establish where their understanding fails.

All assessments offer opportunities for learning; the knowledge gained from such apparently summative tests can be used formatively to understand where further work needs to be done to support the learner's further development of mathematical skills.

 TASK 4.5

What do these answers to the question '1,000,000 − 1 = ?' tell you about the knowledge the learner has about place value?

10000001

9999999

999,999

100,009

Suggested responses to tasks can be found at the end of the book.

PLANNING

Planning is an essential element of the teacher's role. Teachers should plan work based on the initial and diagnostic assessments of their learners and other aspects derived from their ILPs. They also need to consider the whole group while addressing the needs of individuals. Plans should not be set in stone, but always have room for adaptation and personalisation.

Teachers may be required to devise schemes of work that provide an overview of learning activities over a period of time, perhaps for half a term or a term. Teachers are likely to be working towards an accredited assessment and the scheme of work is crucial in allowing the teacher to measure progress towards that final exam, while supporting learners with their specific needs.

The scheme of work may include:

❯ details of formal assessments required

❯ both formal and informal learning objectives

❯ topic content

❯ opportunities for formative assessment

❯ differentiation of learning

❯ identifying a range of types of activity across a programme

See the Appendix for extracts from two schemes of work.

Lesson planning is a crucial skill for all teachers. They need to build a coherent set of learning opportunities from the scheme of work; pitch the content and level of a lesson at the optimum level; build in periods of assessment, instruction, activity and discussion; and include further assessment processes throughout. They need to be aware that some learners will have greater initial knowledge and understanding and will be capable of doing higher-level work than others, and that for some other a great deal of time may need to be spent working through a long history of misconceptions around the topic. The lesson also needs to be interesting, challenging to all and enjoyable!

A lesson plan generally includes:

❯ aims and objectives for the lesson

❯ what the teacher will do and what the learners will do

❯ resources required

❯ opportunities for assessment of learning

❯ notes on how the needs of all learners will be met

❯ timings for different activities

There are two examples of lesson plans in the Appendix.

There are many ways of devising a lesson plan, but in general it involves breaking down activities into sections – time, teacher activity, learner activity, resources used and so on. Try not to focus entirely on the teacher-led activities; ensure you know what you plan for learners to be doing at all times. Clearly your lesson plan will depend upon the learning and teaching methods you choose for the lesson. In turn, the methods you choose will be informed by the group of learners you have and the aims and objectives of the session.

It is generally the case in post-compulsory education that learners will arrive with a wide variety of knowledge and experiences. Your lesson plan should take account of these differences and ensure the lesson works for the entire class; this is termed *differentiation*.

There are many ways of achieving differentiation in teaching, including:

» *differentiation by task*: having more and less challenging tasks. For instance, this may be achieved by having a range of extension activities to provide for learners who complete tasks more quickly than others, or who show greater understanding.

» *open-ended tasks*, which have differentiation built in: learners do as much as they are capable of. Producing posters is an example of an open-ended task that allows learners to work at any level.

» *differentiated groups*: setting different tasks or different timescales for groups with differing abilities. Once you are clear that certain learners have a similar level of understanding of a topic, it is efficient to give that group differentiated tasks and timescales.

» *differentiation by support*: accepting that certain learners will require more support than others. Managing this in a busy classroom may be a challenge, but it is often possible to work closely with individual or small groups while the rest of the class is working on a task.

» *differentiated assessments and feedback*: for some learners, achieving a task to a limited extent may mean they have worked hard, while for others the teacher may expect evidence of greater understanding.

In any one lesson, the teacher may be using any, some or all of these methods according to the needs of the group and the nature of the learning activity.

 TASK 4.6

Take a topic you are planning to teach and plan activities that use the range of different methods of differentiation.

Suggested responses to tasks can be found at the end of the book.

One problem can be that having spent so much time preparing a lesson plan that achieves all the above (and more), teachers can be reluctant to improvise and adapt as the lesson develops. It may be a fine line, but teachers need to recognise the difference between a class being diverted from their main task, and recognising a real opportunity to develop understanding in unexpected ways. Teachers should always aim to be flexible and to use opportunities that arise from learners' own interests.

FEEDBACK AND FEED-FORWARD

Spoken and written feedback is an essential part of the interactive dialogue that takes place between teacher and learners as a central part of teaching and learning. As pointed out above, this discussion, both between learners and with the teacher, gives crucial information about the understanding of learners on the topics they are engaged with. Written feedback, however, tends to be viewed more formally.

Research suggests that the best type of feedback for promoting learning is meaningful comment on the quality of work and constructive advice on how it may be improved, rather than a numerical score.

Helpful feedback should:

> focus on the task, not the mark

> be detailed rather than general

> explain why something is right or wrong

> be concise – don't be tempted to write too much!

> be clear about what has been achieved and what has not

> be timely – feedback should be given as soon as possible to have the maximum impact

> feed forward – it should describe strategies for improvement

Feedback can be given to learners in a number of different ways. Teachers may organise tutorials or one-to-one sessions with learners, comments may be written on work, given orally or perhaps sent as emails.

In a workshop (Race and Pickford, 2007) different types of feedback that students considered to be effective included self-assessing, peer assessment, emailing with attachments, oral feedback to the whole class and constructive questioning within groups. The least effective forms of feedback were marks or grades without comments and individual comments written on work.

Feedback should have an effect on your own teaching as well as the response of your learners. Reflective teachers will consider how the lessons gained from an assessment can affect their own further teaching to that group.

EVALUATION

Reflective teachers should aim to evaluate lessons and longer periods of teaching, as part of a process of continual improvement. They should use evidence provided from learner reactions in the lesson, their responses to activities and their own comments and reflections. The process may be more effective if the teacher uses one of the standard models of reflection to provide an evidence-based evaluation.[1]

Assessment outcomes are an invaluable tool in helping teachers with such reflective evaluations. If learners failed to achieve the outcomes you were planning for, why was that? Try to get into the habit of reflecting on your lessons, using the evidence of learning – or lack of it – that you have assessed along the way. No lesson is ever perfect, but constant reflective practice offers the best way of improving the quality of your lessons.

FINAL SUMMATIVE ASSESSMENTS

For most learners on the majority of courses, the final summative assessment will be the examination.[2] This is used to ascertain whether or not the learner has achieved the aims of the course. In recent years these final assessments have become increasingly more important, being used as an accountability measure for the provider, a benchmark for inspection and frequently success rates have a funding implication.

While educationalists are highly critical about 'teaching to the test', many students appreciate a focus on what they may see as the real reason for taking the course. An approaching examination can be a great motivator, and there are many active learning approaches to revision which can creatively harness that enthusiasm and help your learners achieve. Activities such as deconstructing exam questions or devising marking schemes can capitalise on this motivation and deepen understanding of the assessment process while embedding learning.

Working with learners towards a successful final exam can be done without sacrificing the sustainable learning that is the teacher's goal. Learners who succeed in an examination are likely to increase their confidence, develop their ability to view questions from a number of different perspectives and in many cases increase the range of strategies they can apply. An increase in confidence coupled with the ability to apply a range of strategies will be seen as hallmarks of those who have successfully undertaken a period of mathematical study.

Most mathematics teaching takes place within the context of organisational and administrative systems and an inflexible final examination which is seen as vitally important. Frustrating though these elements may be, imaginative teaching can still provide an exciting, personalised and effective experience for adult learners, building their confidence, knowledge and skill.

---

1. For instance, Gibbs's model for reflection is clear and precise, allowing for description, analysis and evaluation of the experience and helping the reflective practitioner to make sense of experiences, while Johns's model is designed to be carried out through the act of sharing with a colleague or mentor, which enables the experience to become learnt knowledge at a faster rate than reflection alone.
2. At the time of writing, Entry Level final assessments for Functional Mathematics are controlled assessments, set externally but internally assessed.

SUMMARY

This chapter has considered aspects of the teaching and learning cycle. A number of teaching and learning activities have been outlined in other chapters of the book. This chapter has focused on the planning and assessment aspects of teaching with some practical examples in the context of mathematics teaching.

FURTHER READING

Black, P. and Wiliam, D. (1998) *Inside the Black Box: Raising Standards Through Classroom Assessment*. London: NFER Nelson.

DfES (2007) *Thinking Through Mathematics: Strategies for Teaching and Learning*. London: NRDC.

Griffiths, G. and Stone, R. (eds) (2013) *Teaching Adult Numeracy: Principles and Practice*. Maidenhead: Open University Press.

Hodgen, J., Coben D. and Rhodes, V. (2009) *Feedback, Talk and Engaging with Learners. Formative Assessment in Adult Numeracy Research Report*. London: NRDC/King's College London.

Race, P. and Pickford, P. (2007) *Making Teaching Work*. London: Sage.

Swan, M. (2002) 'Dealing with misconceptions in mathematics', in Gates, P. (ed.) *Issues in Mathematics Teaching*. London: Routledge Falmer, pp. 147–65.

5 Teaching practice

INTRODUCTION

This chapter will look at the ways in which you can expect to develop your practice as a mathematics teacher. Some of you may already be teaching and using your job to develop your teaching practice. Some of you will be given teaching practice within an appropriate institution. Whatever the case, it is important to explore the opportunities that you have. Teaching is not just about being in a room with a group of learners, there are also meetings to attend, there is paperwork to complete, new learners to enrol, resources to develop and learner work to mark. Being involved in all these functions as part of your teaching practice will help you in the future.

THE LEARNING ENVIRONMENT

Once you begin teaching you may find yourself in any number of different learning environments. These can range from a classroom in a college to a hall in a community centre or an office in a workplace. The classroom may be equipped with an interactive whiteboard and some individual computers or you may be lucky to have a flipchart or even a wall to attach learning materials to. There may be single desks set out in rows, fold-up tables that you need to set up yourself, one large table or a group of tables placed together in a boardroom style. Don't be afraid of rearranging the room layout to suit your needs if that is possible, but you will need to adapt to whichever situation you find yourself in. This may require you to build up a 'toolbox' of basic materials and resources that you carry around with you.

What materials do you think should go into a basic teaching toolbox?

Suggestions for toolbox:

Mathematics-specific resources:

> Rulers, tape measures, calculators, dice

More general teaching resources:

> Dry wipe markers, flip chart pens or coloured pens, pencils, paper, adhesive for fixing paper to wall, glue sticks, card, scissors, string, post-it notes

You may find other materials helpful but here you have the basic ingredients for a number of mathematics/numeracy activities if you arrive to find the resource cupboard is locked or the computer isn't working. It is not that you will have to spend most of your time improvising with your toolbox but it is always good to be prepared,

TASK 5.1

Design an activity for teaching a mathematics/numeracy concept that learners could engage in using just the materials listed opposite.

Here is one example:

Materials required	Topic	Description of activity
Dice	Multiplication of single digit numbers; factors	A learner throws two dice without allowing others to see the scores. The learner multiplies the scores together and others have to try to guess what the scores were. Once someone has correctly guessed, another learner throws the dice and so on.

Suggested responses to tasks can be found at the end of the book.

As far as possible, you need to try to make sure the learners are comfortable. According to Maslow's hierarchy of needs people require their physiological and safety needs to be attended to before they will be motivated to achieve.

What could you do to try to satisfy the learners' physiological and safety needs?

You may have limited control over some aspects of the environment but you could:

❯ ask learners if they are warm enough or too hot – see if you can adjust room temperature accordingly;

❯ suggest learners bring water into the session;

❯ ensure you plan time for a break – negotiate how long this should be (within reason);

❯ ascertain whether any learners have particular needs that are not being met within the current learning environment;

❯ check that the lighting is sufficient;

❯ ensure those with eyesight or hearing problems are sitting in the best places to see/hear;

❯ check that the room layout is suitable – does it allow for access of all learners and does it suit your planned learning activities?;

❯ warn learners about leaving bags on the floor as they may constitute a trip hazard; and

❯ check that any leads or wiring are secure and not trailing across the floor – this is particularly important where computer equipment is being used.

CLASSROOM MANAGEMENT

If this is your first experience of teaching, one of the most daunting things might be managing a group. You may feel quite confident that you could teach an individual what they need to know about mathematics or numeracy but be unsure about how you can meet the needs of a group of adult learners and keep them all engaged and interested. The key to managing a group is being prepared; you need to have as much information about the learners who make up the group as possible. Once you know who the learners are, what has motivated them to join the class and what their learning goals are, you can plan activities that will engage them and help them achieve their learning objectives.

When you have planned the learning objectives it is good practice to share them with the group. They should be made clear about what they are expected to learn in the session and have an opportunity to ask questions to clarify their understanding. A shared understanding of expected outcomes will assist with managing the learning process. You can share the objectives by showing them in a presentation slide (such as PowerPoint), by writing on a board or flipchart or by printing them out for learners. The latter method is sometimes used where learners are asked to note the lesson objectives on a record of work. It is worth noting that the objectives can be negotiated with learners; indeed, this is considered good practice and will be returned to in Chapter 4: 'The teaching and learning cycle'.

Ground rules are important in sharing expectations – those that learners have of the teacher and those that learners have of each other, in addition to expectations that the teacher might have of the learners. Ground rules work best if they are negotiated and agreed. Rather than ending up with a list of about twenty rules that will not be remembered and followed, it is often a good idea to get learners to prioritise the rules that they feel strongly about and to agree on a short list that they agree are fair and will help the course run more smoothly. You may find it helpful to type up and print out the list so that it can be displayed in the class in order to remind yourself and the learners about what was agreed. However, it is important to apply common sense and flexibility where necessary. Remember that you are teaching adults who have complex lives and there may be valid reasons why a learner does not arrive on time, for example.

 TASK 5.2

What do you think the learners might expect of you as the teacher?

What might they expect of each other?

Suggested responses to tasks can be found at the end of the book.

ICEBREAKERS AND STARTERS

Learning and using the learners' names will help you gain their respect; an icebreaker activity may help you with this.

Sample icebreaker activity: Find someone …

Whose age is an odd number		Who speaks more than one language	
	Whose door number has two digits		Who enjoys cooking
Who enjoyed maths at school		Who has an even number of children	

Each learner is given a 'Find someone' sheet and is encouraged to find a different person's name to write in each box, in answer to the given question. If you take part as well, you will start to learn people's names and find out a little bit about the learners.

It is important to engage the learners as soon as possible and one way of doing this is to provide an interesting starter activity. Starters, also known as warmers, are important activities as they set the scene for the lesson. You can use a starter to give the learners a flavour of the lesson, to assess prior knowledge and/or skills, to pre-teach vocabulary, to recap previous learning or simply to promote interest and enjoyment.

Sample starter activity: I spy a shape

This starter uses the structure of the standard 'I spy' children's game. You will need to decide whether an activity with such connotations will work with your groups – in our experience few groups have a problem if the activity is introduced confidently with some explanation of the purpose. Indeed, you cannot be sure your groups have played such a game, either as or with children, in any case.

Prepare the room by checking out which shapes can be seen. You may need to supplement the range of shapes in the room by using carefully placed photos or objects.

Start the game by saying, 'I spy a shape' and then complete the sentence with either a clue about the shape's properties or give the first letter of the shape's name. For example, 'I spy a shape beginning with the letter C' or 'I spy a shape that has four sides.' Learners try to guess which shape you are referring to and once the shape has been guessed, somebody else says, 'I spy… .' This activity will help you assess what learners already know about shapes and will help you to introduce the topic and related vocabulary in an enjoyable, non-threatening way.

WHAT TYPE OF FEEDBACK SHOULD YOU TRY TO OBTAIN FROM LEARNERS AND HOW?

You should try to find out if the learners enjoyed the session, what they learnt, if they have learnt what they hoped to learn, what they want to follow up, if there were any activities they particularly enjoyed and if there were any they did not. Also ask if they have any suggestions for what could have been done differently.

You can get this information from the learners by asking questions at the end of the session or by getting them to complete a brief questionnaire or a reflective sheet. Alternatively you could use a feedback activity; this could be a fairly straightforward activity such as using coloured post-it notes to 'vote' for the activity they liked the most and the least, followed by a discussion on their reasons and suggestions for change.

DIFFERENT TEACHING PRACTICE FORMATS

There are a number of different formats for teaching practice when you are training to be a mathematics teacher.

A *training class* usually entails a small group of trainee teachers attending the same class and taking turns to teach. If there is a group of four trainees, two of them may teach consecutively in one session and the other two teach in the next session. When they are not teaching, the trainees may be supporting individuals or small groups of learners, or observing the trainee who is doing the teaching. Trainees also get an opportunity to observe the class teacher teaching the group. This format involves the group of trainees planning together under the guidance of the class teacher, who will need to provide the necessary information about the learners. Trainees get informal feedback on their teaching from each other as well as from the class teacher, who may also carry out a formal observation of the trainees when they are teaching.

A *teaching placement* involves a trainee teacher being placed with an experienced mathematics or numeracy teacher – ideally one who has received consistently good grades in their own teaching observations. Unlike the training class, there is usually only one trainee placed in a particular class. The teacher then acts as a mentor to the trainee; there may be a period where the trainee observes the class teacher teaching before the trainee starts to teach the class. The mentor may decide to phase the trainee's practice in by giving them a small group to teach initially or by asking them to teach part of the session, but at some point they will usually progress to teaching the whole class while the teacher observes them and supports as necessary. The mentor will support the trainee with planning and will give them informal feedback on their teaching. They may also carry out formal observations of the trainee teacher.

In-service teaching involves the use of teaching where you are already employed. You may use one of your own groups to demonstrate and develop your teaching practice. In such circumstances it is assumed that your employer will identify a mentor to play a similar role to the mentors mentioned above. In such circumstances, the class is your responsibility and the mentor is likely to spend less time in the group with you, although they will need to observe a number of your sessions. You should agree some regular meetings to discuss planning for the sessions.

If you are already involved in paid teaching and using in-service teaching practice, you will have the advantage of really being responsible for the group and not having to spend unpaid time with a group. On the other hand, you are likely to get less support from your mentor and will have less opportunity to observe others and share ideas, or experience a wider range of learners.

 TASK 5.3

What do you think are the potential advantages and disadvantages of being in a training class or a teaching placement?

Suggested responses to tasks can be found at the end of the book.

The key to overcoming any potential disadvantages of a teaching practice format is to communicate any concerns as soon as possible to your mentor or training class colleagues and to be flexible and compromise where necessary. Being approachable and supportive towards the learners and demonstrating your knowledge and enthusiasm will make it more likely that they will accept you as a teacher rather than a visitor. If you feel you are not getting the opportunity to observe different teaching styles you could arrange to visit other classes and observe the teaching. This is a good idea in any case as you can arrange to observe teaching in different contexts and at different levels.

MAKING FULL USE OF TEACHING PRACTICE AND WORKING WITH A MENTOR

If you are allocated teaching practice it will be for a relatively short period of time and therefore you will need to make the most of the opportunity. You will have the chance to experience support with your planning, observe an experienced teacher, try out ideas in a supported environment and receive informal supportive feedback on how you are progressing. If you are placed in a training class rather than a placement, you will also get the opportunity to observe your peers trying out different approaches, learning materials and activities. This type of support is unlikely to occur again in your teaching career and you should learn a considerable amount in a relatively small timescale. You will probably be assisted with finding a mentor but if you are able to get involved with this process, try to ensure that the mentor is experienced and enthusiastic and is able to dedicate sufficient time to the mentoring process.

So how can you extract the most from this experience?

❱ Try to observe the class teacher as much as possible.

❱ You may be asked to support other learners while the teacher is teaching but there will be times when this support is not required. Use the time to watch how the teacher manages the group and individuals, how they start the session, how they interact with the learners, how they assess learning and feed back on learner progress and how they end the session.

❱ Ask for all the relevant documentation, such as the scheme of work and lesson plans, and agree a time when you can discuss the planning process and how the teacher felt that the lesson went.

❱ Watch the learners and how they react to the teacher.

❱ Watch how they respond to questions and the learning activities.

❱ Ask yourself if they are enjoying learning and if they seem to be making progress. Do there appear to be any issues? If so, why have they occurred?

❱ Write a reflection on what you have observed, recording what worked well and anything that didn't, with possible reasons. Try to be objective – even experienced teachers will do things that do not work so well for various reasons and you can learn from these incidents just as much as from observing things that work well.

Once you start teaching, aim to get the most out of joint planning by being prepared. Try to obtain information about the individual learners and get to know them as much as possible. Come with your own ideas and don't be afraid to suggest something new or different while taking the guidance of your placement teacher or mentor on board. Look for and collect potential teaching materials at every opportunity, such as newspapers, leaflets, websites and so on. Reflect on any theory that you have been taught or literature that you have read and think how it might apply to planning for the group. If you are in a training class, discuss the upcoming session together and spark ideas off each other for activities and materials. Discuss particular learners with each other and how you can support them most effectively.

The feedback process is immensely important in your development as a new teacher. You will be observed throughout your teaching career and should receive constructive feedback at all times. Observations during your teaching practice are intended to be supportive and developmental, rather than being judgemental, like those during inspections. You will get the opportunity to see yourself through different eyes, i.e. those of your mentor and, if you are in training class, those of your colleagues. When you first start teaching there may be a number of points for development – this is to be expected. If you are receiving critical feedback from your peers you might agree on feeding back on one strength and one area for development. Take the suggestions for improvement that are offered (if they are not offered, ask for suggestions) and incorporate them into your own reflections on the lesson. You should also aim to get feedback from the learners. Avoid just asking them if they have learnt something or if they enjoyed it – both of these are important, but you should aim to obtain more detailed feedback to inform your own reflections.

BELIEFS AND TEACHING APPROACHES

Does the traditional, transmission-style approach outlined in the introduction sound similar to your own experience of being taught mathematics? How enjoyable did you find that approach? Did you feel that approach worked for you and others in your class? Do you feel your own experience will shape the way you teach mathematics or numeracy?

Research has found that many teachers have a set of beliefs about mathematics, which may be connected to the way they were taught, and this belief system often translates into teaching in a particular style. For example, teachers who believe that mathematics is a fixed body of knowledge may lean towards wanting to transmit their knowledge to the learners, i.e. follow a predominantly transmission teaching style. On the other hand, teachers who believe that mathematics can be used in different ways to solve different problems that have multiple solutions may be more disposed towards trying to make the learning relevant to their learners' lives and experiences, involving the learners more directly in the process. However, there are exceptions to this trend and it is important to be open to trying different approaches. You may have enjoyed mathematics at school and achieved success whether you were taught in a traditional way or not, but consider whether the same could be said for your classmates. Remember, too, that adult mathematics learners, in general, have been through an education system which has not been successful and enabled them to achieve their potential in the subject. They often come with anxieties and attitudes towards mathematics that stem from their earlier experiences of being taught the subject and this should be considered when deciding which teaching approaches to use.

SUMMARY

We hope this chapter has demonstrated the importance of teaching practice in training to be a teacher. There are likely to be some challenges in undertaking the opportunities that you are presented with but overall this will be a positive experience and one from which you will gain much learning.

FURTHER READING

Griffiths, G. and Stone, R. (eds) (2013) *Teaching Adult Numeracy: Principles and Practice.* Maidenhead: Open University Press.

Swain, J., Newmarch, B. and Gormley, O. (2007) *Numeracy. Developing Adult Teaching and Learning: Practitioner Guides.* Leicester: NIACE.

6 Examples of teaching activities

INTRODUCTION

In this chapter we provide examples of a range of different types of activities indicative of those expected within the sector. This is not to suggest that you should never use standard worksheets, or indeed deliver presentations, but you will be expected to use a range of activities in every session you teach. To help you employ a range of activities you can build up a bank of your own ideas and resources as well as using any available, ready-made resources (see the links at the end of the book for places to find resources). However, it is important that the activities are appropriate for the group of learners and for the aims of the session. This may require you to adapt the activities for different groups and situations.

STARTER ACTIVITIES

It is recommended that you use a starter activity at the beginning of each session. These can be used to recap previous learning or as an introduction to the main session.

Resource-lite starters

You can play a twenty-questions-style starter for any topic. Start off using an example yourself – think of a mathematical concept or item for the focus of the activity.

A simple example is to have a number in mind and learners try to work out which number you are thinking of by asking questions. More complex examples of this activity could involve a range of different mathematical ideas. If you want to focus on geometry, you could think of a 2D or 3D shape. Learners ask questions to which you can only answer yes or no. For example: Are you thinking of a rule? Are you thinking of a type of angle? Are you thinking of a 2D shape? If you are asked more general questions, remind them that you are only allowed to answer yes or no.

Bear in mind the terminology that may be required for such a task. The geometry example would almost certainly need to follow sessions that have addressed a range of geometrical ideas so that the learners are able to use and understand the relevant vocabulary.

Follow me chains

A common starter involves a chain of cards where learners are each given a card. Each card contains an answer to another's question, together with its own question. The cards are shuffled and distributed (or you might target easier/harder cards to particular individuals). Then the chain starts – you might begin with a card of your own, carefully read out the question and point out that someone has the card. If no one responds, ask for all learners to think what the answer might be to help out. At the beginning you will need to progress the chain carefully but after a few goes you will be able to reduce the support.

Be aware that such chains need all cards to be used, so you may have to give some learners more than one card. Note that the question on the final card has the answer on the first card, thereby completing the chain.

There are card sets that can be bought, although it may be easier to target your own group by producing your own.

Example cards with a focus on measurement conversions.			
I have 50 cm Who has 70 mm?	I have 7 cm Who has 2 m?	I have 200 cm Who has 60 mm?	I have 6 cm Who has 150 cm?
I have 1.5 m? Who has 75 mm?	I have 7.5 cm Who has 350 mm?	I have 35 cm Who has 250 cm?	I have 2.5 m Who has 3.5 cm?
I have 35 mm Who has 25 m?	I have 2500 cm Who has 750 mm?	I have 75 cm Who has 750 mm?	I have 0.75 m Who has 500 mm?

Open questioning

You could ask learners to write down as many things that they know about a mathematical idea, for example probability. Such open tasks can allow most learners to respond in some form. Think about how you will express the task, as the task itself is likely to produce different results.

For example, 'Write down as many things about probability that you remember' will most likely produce more items than 'What does probability mean?'

 TASK 6.1

Research and/or develop further examples of starter activities.

Suggested responses to tasks can be found at the end of the book.

DISCUSSION AND COLLABORATION ACTIVITIES

The first two activities in this category make use of mini whiteboards (or laminated sheets of paper if these are not available). This resource enables the tutor to obtain a lot of feedback from learners at the same time and to assess their knowledge. In these examples, their use also provides the learners with a set of information that they can use. This has the dual effect of reducing the amount of preparation that the tutor has to do and ensures that the information is relevant to the learners as it comes directly from them.

The first two activities are tutor-led, with active learner involvement; therefore the instructions are aimed at the tutor.

Guess the average activity

Instructions:

> Each learner is given a mini whiteboard or laminated sheet, a dry wipe marker and a cloth or sponge.

> The tutor poses questions such as, 'How long did it take you to get to college?' or 'How much is a loaf of bread?' and learners write the answers on their board/sheet. The tutor could also ask learners to pose their own questions.

> Learners then show their answers and the tutor asks them to guess the average answer. The type of average could be specified or it could be left open at this point. This is followed by questions on how they have guessed the answer, which would include a discussion on different types of average.

> The tutor records the different answers and asks learners to calculate the average (specified or all types).

The discussion could be extended to look at simple spread of data (range).

Guess the fraction activity

Instructions:

> Each learner is given a mini whiteboard or laminated sheet, a dry wipe marker and a cloth or sponge.

> The tutor poses questions such as, 'Do you have a pet?' or 'Do you prefer reading or listening to music?' Learners can suggest their own questions.

> Before they show their answers, the tutor chooses someone to guess the fraction (or decimal or percentage) who have answered in a certain way. Everyone else then shows their answers and the learner checks the results against their guess. The tutor goes through the process of calculating the fraction (or decimal or percentage, as appropriate) if anyone is unsure and could also ask learners how they could convert the answer to a different form (for example, fraction to decimal). Discussion can be extended to include, 'What is the ratio of ….. to …..?' and 'How could you present the information in a diagram?'

Factorising and expanding quadratic expressions activity

This is an activity that can be used when teaching higher-level GCSE. It is a collaborative activity that encourages peer support and assessment. This activity is learner-led and the instructions are aimed at the learner. The tutor's role here would be to facilitate the activity and provide guidance. Instructions:

❯ Working in pairs, each learner writes down a pair of brackets, each bracket containing a letter and a number (the number may be positive or negative but use the same letter in both brackets). Do not show what you have written to your partner.

❯ Expand the brackets to give an expression.

❯ Simplify the expression.

❯ Pass the simplified expression to the other learner and ask them to factorise it by putting it into two brackets.

❯ Check that this matches the original brackets – if not, try to work out where an error has been made.

Measuring and estimating practical activities

The next activity is a practical activity involving measures. This topic requires practical work to help learners understand the concepts involved. Here, learners are practising their estimation and measuring skills relating to length and weight. The initial measuring activity provides the comparison which assists with the estimation task. Doing the activity in groups encourages the learners to discuss likely measurements and the units involved. Comparison between the estimates and actual measurements allows for some peer assessment and further learning. The instructions below are aimed at the teacher although they could easily be adapted to give to the learners.

Length

❯ Learners work in small groups (maximum three per group).

❯ Each group is given different coloured post-it notes.

❯ The teacher introduces the activity by getting the groups to accurately measure the dimensions of a given object. The teacher checks the results and discusses any discrepancies.

❯ Various different objects are placed around the room. The groups visit each object and guess the length or width (without measuring), writing their answer on a post-it note which they place beside the object.

❯ Once they have guessed the required dimensions of all the objects, the teacher takes each object in turn and allocates a group to accurately measure it. Comparisons are then made between the accurate measurement and the estimates.

Weight

Repeat steps above by estimating weight and weighing a set of objects.

Fill the gaps (practise in word/context problems)

Success in examinations usually requires plenty of practise in answering the type of questions that will be encountered in the test. Learners often practise/prepare for exams individually, but this activity is an example of a collaborative activity designed to help learners prepare for answering contextualised word problems – as in functional mathematics assessments. The small-group approach encourages discussion and sharing of ideas.

In this type of activity the tutor selects an appropriate word problem and removes key numerical content. The *fill the gaps* approach encourages the learners to get more involved with the question by inserting their own information, which may help them to have a better understanding of the problem. Here are instructions for such an activity.

❱ Remove the numerical information from a suitable contextualised word problem. Add instructions directing the learners as to the type of information that should be added and where.

❱ Learners should work in pairs or groups of three to decide on the information that is missing from the functional skills problem and to complete the question.

❱ This should be followed by whole-group discussion on the problem and display of their results.

Example (adapted from Edexcel Functional Skills Level 1 paper July 2012)

There is a sports tournament between the following teams (insert team names):

…………………………...

…………………………...

…………………………...

…………………………...

Each team will play all the other teams once.

The first game starts at ………………………… (insert time)

Each game takes …………………………… (insert length of game and units of time)

There is a break of ……………………….. (insert length of break and units of time) after each match.

Only one game is played at a time.

Produce a timetable for the tournament showing the teams playing in each game and start and end times of the different games.

ACTIVITIES INVOLVING IT

Information technology has an important part to play in education. Most learners have some experience of using technology and it can add interest and increase accessibility. It can be incorporated into activities in many different ways, depending on available facilities.

Activity on positive and negative numbers

Instructions:

❯ Learners work in pairs.

❯ Give learners a country to research (using a smartphone, tablet or PC). Ask them to find out the average temperature in January for that country. (Ensure that you include some countries with average temperatures above 0°C and some below 0°C.)

❯ Display a large, graduated, but unlabelled (apart from 0°C) number line on a smartboard or on a wall – a vertical number line is preferable.

❯ Select two of the countries and ask learners to mark their country's average temperature on the number line. Ask them to explain how they worked out where to make the mark and how they can check if it is in the correct position.

❯ Ask questions about the difference between the two countries' average temperatures, for example:

- What is the difference between the temperature in A and B?
- How many degrees warmer is A than B (as appropriate)?
- How much colder is B than C (as appropriate)?

❯ Ask them to demonstrate how they worked out their answers (using the number line to demonstrate).

❯ Repeat with two other countries and so on.

Shapes poster

The following activity can be used on a Mathematics and ICT course.
The activity would be staged after some input on 2D shape and has the dual purpose of assessing knowledge of 2D shape names and providing practice in word processing. Learners work on the activity individually but could be given the option of using peer support.

Instructions (aimed at the learners):

❯ Open a blank Word document.

❯ Insert different shapes on the page using 'Insert, Shapes'.

❯ You can change the size (click on the shape and click and drag one of the spots).

❯ You can change the colour of the shapes (click on the shape and use 'Format, Shape Fill').

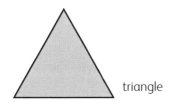

triangle

❭ Draw a text box next to each shape ('Insert, Text Box, Draw Text Box').

❭ Write the name of the shape in the text box.

❭ Add a page border: 'Page Layout, Page Borders'.

❭ You can try out different borders, e.g. try art.

❭ Apply to whole document.

❭ Save and print your document.

INVESTIGATIONS

Investigations allow for the practising of a range of skills and can usually be accessed in different ways by learners working at different levels.

Making up amounts of money

The following investigation is an interesting way of exploring number but also may have practical uses concerning the selection of change when shopping. It is known that some individuals are so uncomfortable with choosing the right amount of money because they may get it wrong, that they always round up. This means that they acquire a large amount of change that they are unsure how to use.

Use combinations of 10p, 50p and £1 coins.

How many different ways can you make £1.50?

Which way(s) uses the most number of coins? Which way(s) uses the least number of coins?

Now do the same for …

❭ £1.80

❭ £2.40

❭ £3.00

❭ £4.50

Tiling pattern

You have square tiles of two different colours – in this case dark and light.
Here are two different repeating patterns.

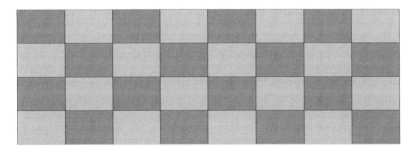

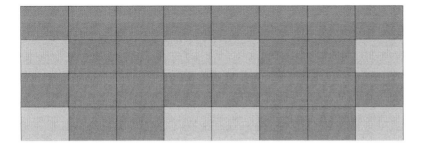

Investigate how many patterns are possible.

The amount of guidance you give will depend upon the group – and subgroups – involved. Do you use repeated patterns of four tiles or widen the possibilities? Do you have a fixed wall size? Do you encourage the use of three or more different tiles to expand the investigation?

ACTIVITIES INVOLVING LANGUAGE AND CULTURE

The following activity is designed to make explicit the language structure that is used in the standard way of telling time in the UK. This is something which varies considerably between languages and cultures.

How do you tell time in your language?

A blog from a Slovak speaker notes the different ways of telling time in some languages. Some examples that can be seen at http://www.flash-sticks.com/blog/different-country-different-ways-to-tell-time/#.Usn0tPsQRbw:

❯ In German and Slovak, a typical way of expressing a time like 2:30pm would be to say that it is 'half 3' – 'halb drei'; 'pol tretej'.

❯ In German, the literal meaning of 'Dreiviertel drei', meaning it is 2:45pm, would be 'three quarters three'.

❱ French for 2:45pm, 'Il est deux heures moins le quart', literally means it's two minus a quarter.

❱ The Spanish 'son las tres menos cuarto' means it's three minus a quarter.

❱ The Italian 'sono le sei meno un quarto' means it's six without a quarter.

Activity

What ways of telling time are there in languages you know?

The next activity uses the idea of a well-known game (i.e. Pictionary) to investigate the way in which some words have a number of different meanings – particularly in their mathematical and non-mathematical use.

Everyday and mathematical words Pictionary

There are many words that have different meanings in mathematics than in the everyday. For example, table – in mathematics it is a way of displaying information, in the everyday it is a piece of furniture. Look at the words in the following and draw/sketch something to explain the different meanings.

Word	Everyday meaning	Mathematical meaning
Table		
Difference		
Mean		
Factor		
Work out		
Digit		
Scale		

Can you think of others?

❭ There are words that have similar but slightly different meanings in mathematics to the everyday. For example, perimeter which can mean the physical description around a shape (a perimeter fence) in the everyday, while in mathematics we mean the length.

❭ Range in the everyday is usually thought of as from … to …, whereas in mathematics it is a single number.

❭ Possible usually means possible but unlikely in the everyday, whereas in mathematics it can span unlikely to certain.

Can you think of others?

Teacher's notes:

❭ The purpose of the activity is to identify words which have more than one meaning and enable discussion of these.

❭ Some words are rarely mistaken (for example, table is not often confused) but can be used to illustrate the point and give some learners a chance to participate.

❭ It is recommended that the sheet is used as a way of recording results rather than as a worksheet to be completed. There are a number of ways that the activity can be undertaken. Learners could be split into teams working on one type of meaning only and the words are selected randomly from a set of cards.

❭ Learners could work in small groups on one word, finding both meanings using a prompt sheet.

❭ Learners could take one meaning each to work on, using prompt sheets

Some other words you could use are: row, column, odd and even.

GAMES

Games can provide a fun, non-threatening way of learning maths. They do not need to be purchased and do not need to be complicated or take hours to produce. Here are some examples of games that can be used to support learning.

Place value tables activity

Instructions:

❭ Ask learners what place value means – you may need to prompt them with the example of tens and units.

❭ Ask them to give you the place value column headings – talk about the value increasing as you move to the left.

❭ Hand out place value table sheets (alternatively, hand out blank tables and get the learners to write in the headings).

Learners can do the activity individually or in pairs or threes. Inform them that you will be calling out digits from zero to nine and they should write them in the place value table to create the biggest number they can.

❭ Throw a ten-sided die and read out the number. Instruct learners to write the number in a column of their choice but warn them that once they have written it down they cannot change its position.

❭ Repeat by throwing the die and reading out the numbers until the learners have a digit in each column.

❭ Ask learners to read out their numbers and record them on a flip chart. Allow learners to just say the digits if they are unable to read out the whole number in words. If some learners are having trouble saying the number in words, go over strategies for reading large numbers (such as grouping in threes or using a place value grid). Ask learners for their strategies.

❭ Compare the numbers created and see who has the largest/smallest number.

❭ Ask learners to reorder the different numbers created so that they are in order of size (ascending order).

❭ Discuss how they did this.

❭ Repeat the activity, asking learners to make the lowest number they can. Discuss strategies for success.

Extension activity:

Learners who are able to work with decimals can be given the table with the decimal place values included. They would need to work in a separate group, competing against each other to make the highest or lowest number as directed.

Temperature game

Instructions:

The game may be played with two players, or four players in two teams. Each player or team has a set of counters – use a different colour for different players or teams.

❭ Players throw a die to see who goes first. Player 1 places their counter on the 0, then tosses a coin to see if they should add or subtract (heads = add, tails = subtract). They toss the coin a second time to see if the number to be added or subtracted is positive or negative. Then throw a dice to get the value of the number.

❭ They then work out which direction to move and move the number of spaces given by the score on the die. For example, Head, Tail, 3 means they should add (-3). From a starting position of 0 this will take them down to -3 on the number line.

❭ Player 2 then places their counter on 0 and goes through the same process.

❭ This continues with the players taking turns until a player falls off the end of the number line and loses their counter.

If more practice is needed, the player/team who have lost their counter may start again with a second counter and so on.

Players are encouraged to try to work out their moves with the analogy of adding or taking away heat or ice helping them to understand the abstract concept of adding and subtracting negative numbers.

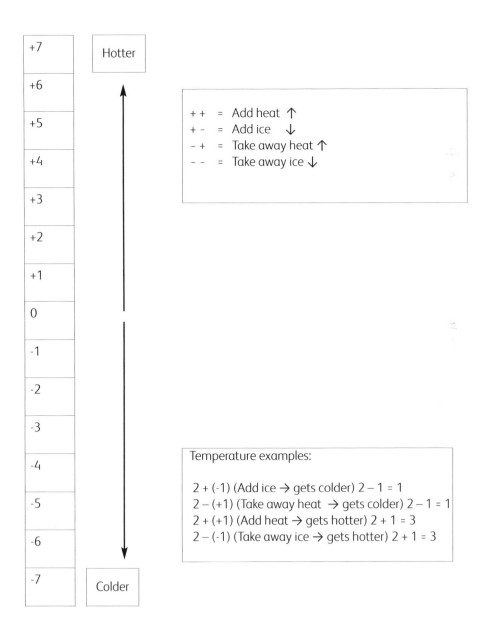

+7

Hotter

+6

+5

+4

+3

+2

+1

0

-1

-2

-3

-4

-5

-6

-7

Colder

```
+ +  =  Add heat ↑
+ -  =  Add ice  ↓
- +  =  Take away heat ↑
- -  =  Take away ice ↓
```

Temperature examples:

2 + (-1) (Add ice → gets colder) 2 – 1 = 1
2 – (+1) (Take away heat → gets colder) 2 – 1 = 1
2 + (+1) (Add heat → gets hotter) 2 + 1 = 3
2 – (-1) (Take away ice → gets hotter) 2 + 1 = 3

Algebra game

Start →

Solve it

Solve it

Solve it

Solve it

Solve it

Solve it

Solve it

Solve it

Each player places a counter on the start.

Take turns:
- Take an expression card and throw a dice. Substitute the score on the dice for x and move that number of spaces.

- If you land on a 'Solve it' square take a 'Solve it' card and work out the value of x then move that number of spaces.

- The winner is the first one to go round the board twice.

Cut out and shuffle the expression cards. Place in a bag or envelope.

x	$2x$	$3x$
$4x$	$x+1$	$x+2$
$x+3$	$x+4$	$x+5$
$2x+1$	$2x-1$	x^2
x^2-1	$3x+1$	$3x-1$
$3x-2$	$3x-3$	$2x+2$

'Solve it' cards – cut out and place face down on board.

$x+1=5$	$2x+1=5$	$x-1=2$
$x^2=36$	$2x+2=4$	$3x=15$
$x/2=3$	$3x+1=10$	$x-3=4$
$x-2=5$	$2x-1=5$	$4x=16$

TASK 6.2

This section has introduced examples of activities under the following headings:

❭ Starter activities

❭ Discussion and collaborative activities

❭ Activities involving IT

❭ Investigations

❭ Activities involving language and culture

❭ Games

How do these relate to the educational theories and teaching approaches discussed in Chapter 3?

Suggested responses to tasks can be found at the end of the book.

SUMMARY

This chapter has introduced a number of examples of activities that can be used with learners. Teachers should be aware that they are expected to research and develop activities for their own groups. The balance of how much work is involved in researching existing resources as opposed to adapting or developing your own will depend upon a number of factors. These factors include the availability of resources for the particular mathematical topic, the needs of the groups and your own skills of development. It is also worth noting that the relationship between activities, educational theories and approaches to teaching are not straightforward, one-to-one mappings. As such, your rationale for choosing an activity may draw upon a range of issues and ideas.

FURTHER READING

Bouch D. and Ness C. (2007) *Measurement*. London: NRDC. At: http://www.nrdc.org.uk/publications_details.asp?ID=97

Bouch, D. and Ness, C. (2007) *Time and Money*. London: NRDC. At: http://www.nrdc.org.uk/publications_details.asp?ID=99

Bouch, D. and Ness, C. (2007) *Topic-Based Teaching*. London: NRDC. At: http://www.nrdc.org.uk/publications_details.asp?ID=98

McLeod, R. and Newmarch, B. (2006) *Fractions*. London: NRDC. At: http://www.ncetm.org.uk/files/257666/fractions_booklet.pdf

Newmarch, B. and Part, T. (2007) *Number*. London: NRDC. At http://www.ncetm.org.uk/files/257643/number.pdf

7 Planning for inclusive practice

INTRODUCTION

An important consideration when planning courses is the extent to which all learners feel included. There are many reasons why an individual may be or feel excluded from a session. By noting what might go wrong it may help to plan and be prepared. Before considering a number of cases, it is worth noting a general principle which is proposed around inclusion: to avoid seeing individuals (or groups) as being deficient but, rather, to consider how a course or activities can make things difficult for individuals. It is also worth noting that the discussions below identify some groups of people. In developing ideas and moving education forward, it has been helpful to use such groups, and indeed some individuals for various reasons want to be seen to be part of a group, but there is a danger here of seeing these groups as 'other', not part of 'us'. This is not the case. The fact is that the discussions involved here are about all of us. Indeed, one of the messages we wish to emphasise is to avoid stereotyping and note that we all have different knowledge and skills that we bring to education.

SOCIAL GROUPS

There are a number of social groups that have traditionally been considered in relation to education – we identified some of these in Chapter 3. *Gender* is a key one. We noted earlier that mathematics has traditionally been seen as a male subject. As much in society has changed, this is a far less prevalent view than in previous times. Nevertheless, some studies (in this case with A-level learners) have shown that some attitudes appear to have a gender specific impact. Heather Mendick notes some binary oppositions that were evidenced within her research and implied a gendered stereotyping. These included:

> Numbers/words

> Ordered and rule based/creative and emotional

> Fast/slow

> Competition/collaboration

> Independent/dependant

> Active/passive

> Naturally able/hardworking

> Active/passive

In these oppositions, the first are seen as associated with males while the second are linked to females. Mendick notes that this language is embedded within culture and argues that it is important to counter this language.

Ethnicity and cultural groupings are another area where there is a risk of stereotyping and potential societal self-replication. For example, data from the 1990s reported that groups of Asian school children were more positive about mathematics than white, European groups. There is a danger that self-fulfilling prophecies are being met through differing expectations of various ethnic groups.

Another key grouping is *social class*. Some key studies have noted that there is a strong link between education levels and social class. If one looks at the proportions of eligible individuals who attend post-compulsory education in each social group, it is clear that lower socio-economic groups, while having larger absolute numbers attending, do so in a lower proportion than those in higher socio-economic groups. This has led many to see society as essentially replicating social inequalities. Indeed, some have investigated what this might mean at the level of teaching. Paul Dowling studied a series of school mathematics text books. He argued that the 'higher' level texts positioned the reader quite differently from the lower-level texts. He provided examples to illustrate this difference. In two parallel contexts looking at road traffic accidents, he noted how the higher-level text presented the scenario from above, as if positioning the reader as an authority over the scene. In contrast, the lower-level text positioned the reader at the elbow of a policeman as if the reader was taking part in the scenario.

Whether we consider gender, ethnicity or social class, the contexts and examples that we choose to illustrate mathematics may have some impact on expectations and the extent to which we are part of a self-replicating system. Of course, the contexts and examples cannot suit everybody, all of the time, but we could certainly make a decision to include a range of examples – perhaps sometimes taking ideas outside of the expectations you may have for a group.

 TASK 7.1

Identify a range of contexts that involve measurement and consider what stereotyping might occur.

Suggested responses to tasks can be found at the end of the book.

LEARNING DIFFICULTIES AND DISABILITIES

Over time, educationalists have identified that individuals process information in different ways. Some groups of individuals have been noticed struggling with traditional, transmission-style educational approaches and have been identified as having a learning difficulty or disability. It should be noted that most of these conditions are expressed in terms that suggest a deficit model. There is a move to re-express such deficiencies as diversity. The range of issues raised have been associated with the way the brain perceives the world (rather than as a physical deficiency) and so some writers have begun to talk about *neurodiversity*. It is also worth noting that the term 'disability' is being seen positively by some groups.

Learning difficulties are understood as ways in which the brain either interprets information from the world or responds to the information differently than might be understood by the average individual.

Some key terms are:

❭ *specific learning difficulties and disabilities* such as *dyslexia, dyspraxia,* and *dyscalculia*. For dyslexia, the brain interprets what it sees or hears differently to what might be expected. For dyspraxia, the body is not controlled in ways expected so that writing and physical actions may be seen as 'clumsy'. Dyscalculia is not well understood but it is thought that some individuals conceive of numbers differently from the majority. They are known as specific learning difficulties because they tend to affect learning of some subjects and areas more than others.

❭ *mild, medium and severe general learning difficulties and disabilities*. Individuals with general learning difficulties find learning difficult in all or most subject areas.

❭ *autistic spectrum disorders*. Some individuals appear to struggle with emotional responses and demonstrate a lack of empathy with others.

These are all very different types of condition that will affect people differently. In all cases you will need to consider the effect of giving extra thinking time and reducing pressure on learners. Avoid too many directed questions that don't have a good amount of thinking time attached.

We have noted that traditional, transmission-style approaches to teaching mathematics have tended to develop mathematical ideas step by step, building mathematics up. In some ways this works well with learners on the autistic spectrum. It is worth noting that it helps to write down these steps and to be careful about using idiomatic expressions (such as, 'I wonder if ').

On the other hand, those with dyslexia often lose where they are in the detail. Such individuals prefer to see the bigger picture first before delving into detail. In addition, it may be important to try a number of different ways of undertaking mathematical processes to find the idea that fits with the learner's thinking.

The publication *Access for All* (BSA, 2001) outlines a number ideas of how to support learners. For example:

❭ use visual images, including colour, and practical equipment to model and demonstrate conceptual ideas;

❭ encourage learners to explain their thinking;

❭ encourage learners to make notes in ways that assist them, such as 'mind map'-style diagrams.

It is worth noting that it is possible that some learners are not able to process information in the same way as others. For example, one notion of dyscalculia would suggest that some individuals find it difficult to subitise – the ability to recognise small amounts of objects without having to count. Other ideas relate to the difficulty – or delay – in recognising the meaning of numerals (i.e. how they relate to objects). Practice will not improve this for such individuals, so a teacher will need to look for alternative strategies and aids that stop this being an issue when undertaking tasks.

Overall, we would advise talking to learners, finding out what works for them and trying to incorporate these for the whole class but bearing in mind that some approaches may be less successful with others.

PHYSICAL AND MENTAL CONDITIONS

Similar to the move from deficit models of learning difficulties, there has been a move from seeing physical and mental disabilities via a 'medical model' of deficiency to what is known as the *social model*. In the social model of disability, the environment is seen as providing the difficulty rather than the disability.

For learners who are *partially sighted* it may be a case of rethinking activities and delivering differently to the whole group, rather than necessarily providing separate written information.

Learners who are *deaf* or have *partial hearing* may be entitled to, or bring their own, signers or transcribers. Even so, it is worth realising that sign language does not have equivalents of certain concepts in spoken languages (for example big, bigger and biggest are usually signed the same way). Also, the signer may not have strong mathematical skills themselves and may communicate inaccurate messages. However, by rethinking activities to induct meaning you may well help all learners.

 TASK 7.2

Identify some ways in which you might plan for learners with specific difficulties or disabilities.

Suggested responses to tasks can be found at the end of the book.

LANGUAGE AND MATHEMATICS

Language plays an important part in learning. We have noted the significance of collaborative learning in the section on approaches to teaching and learning mathematics. Whether a learner is a native speaker of English or not, learning will involve the *four language skills* of speaking, listening, reading and writing. Learners returning to education may also need their *literacy* developing, and the same may be true for speakers of other languages (usually identified as *ESOL* learners – English for Speakers of Other Languages); it is therefore important that teachers should consider the ways in which they communicate. There are many things to consider when looking at language and mathematics and we would like to identify a few which should help you at the start of your teaching.

Written English in 'word problems' can cause some particular issues. Consider the word 'charabanc'. Does this mean anything to you? This is a bit of an obscure word – and not particularly related to mathematics. Nevertheless, it raises some issues to consider and is included because it will put some readers in the place of a learner. In the UK, 'charabanc' was a term used to describe a type of open-topped coach used by holiday makers but has now become fairly obsolete. It is quite likely that some groups of older learners will be very familiar with the term, while others have no knowledge. And of course, speakers of other languages are unlikely to have come across it. And how would you pronounce it? The word actually derives from a French term – char-à-bancs, cart with seats – and like many words in the English language may not be obvious how it is pronounced. In the case of charabanc, there is a soft 'ch' sound at the beginning, like in 'charade' rather than 'chalk', and usually the end sounds like 'bang' (although some pronounce the word more like 'bank').

You may use words that you expect to be *familiar* to learners that may be *obscure* to them – for example, the word 'vertical'. You may use *contexts* for questions that involve non-mathematical words that may not be understood. For example, a recent national mathematics assessment used information about the Lake District and 'steamer' transport. The examiner's report noted that many candidates appeared to misunderstand what a 'steamer' was and also noted that the question was badly answered overall. The suggestion here is that meaning was lost for many participants. This one word made it difficult for candidates to follow through with the responses expected – even when it did not appear to matter whether the candidates realised what a 'steamer' was.

It is important to encourage learners to ask questions about the meanings of words – and you can encourage this by taking time to ask about the meanings of words in activities.

Learners may be unsure of how to *pronounce* words – and may not be able to link what you say to what they see on the page. If you use the word 'isosceles' in your spoken English and draw triangles but never write the word, there is a danger that learners will not link the written and spoken words. Indeed, they may not 'hear' the component parts of words and create their own version of the word in their heads. So do take the time to write up key words as you speak them and build in activities that involve learners speaking and listening as well as reading and writing. For example, you could pair up learners to ask and answer sets of questions with each other that use terminology you have introduced.

There are words that have *more than one meaning* – including ones for which the non-mathematical use is more significant for people. Take the word 'product'. In the mathematical context, you might ask 'what is the product of 3 and 5?' This is a relatively obscure use of the word for most who probably will not associate this with 'multiplication'. The word 'sum' is usually better understood – although not always associated with 'addition' and sometimes seen as a general term for all calculations. The word 'range' in statistics has a slightly different meaning in mathematics than in general use. In mathematics we are looking for a single number as an answer (biggest–smallest), whereas in general use the word tends to imply the two extreme values (from x to y). This suggests that there is a point to taking time to discuss different meanings of words (for example see the mathematical Pictionary activity in Chapter 6).

TASK 7.3

What language and context issues do the following words raise when used in a numeracy session?

a. Mode

b. Modal

c. Fifth

d. Find

e. Work out

f. Kezban

g. Possible

h. Risk

i. Circle

j. Football

k. Cooking

l. Dinner

Suggested responses to tasks can be found at the end of the book.

(This task is based upon a similar activity produced for the Skills for Life Quality Initiative Level 4 professional development modules for adult numeracy teachers.)

The issues raised so far are relevant to native speakers as well as ESOL learners. The final points we would like to raise in this introductory text are two issues that differentiate these groups.

One thing to bear in mind is that many ESOL learners have a strong mathematical background in their own language but will fail to achieve UK qualifications through their need to develop their English skills. In such cases, the job of the teacher becomes one of supporting English development using their knowledge of mathematics. Our final point concerns how a teacher responds when they become aware that a learner is unclear on what to do with a task. A natural thing to do when teaching in your own language is to re-express the task using alternative words and phrases. This can be very effective with native speakers but can be more confusing for ESOL learners. Indeed, it may be that they need the task to be repeated so that they have understood all the parts.

SUMMARY

This chapter presented a number of issues that you should take into account in seeking to include all learners in your teaching sessions. There are some important things to think about when choosing contexts for the mathematics that you work with so that you use examples that interest while avoiding stereotypes. There are issues in how you represent tasks – how much is verbal and how much written – and how clear instructions are provided. Overall, the message is to listen to learners about their interests and their strengths and to build in time to discuss how they best learn. You probably can't satisfy all learners' needs at all times but over time you can explicitly address each of them.

FURTHER READING

DfES (2002) *Access for All: Guidance on Making the Adult Literacy and Numeracy Core Curricula Accessible*. London: DfES.

Griffiths, G. and Stone, R. (eds) (2013) *Teaching Adult Numeracy: Principles and Practice*. Maidenhead: Open University Press.

Lee, C. (2006) *Language for Learning Mathematics. Assessment for Learning in Practice*. Maidenhead and NY: Oxford University Press.

8 Professionalism

INTRODUCTION

What factors are important in being a professional teacher of adult mathematics? There are a range of sources that might help with the answer to this question. We could ask other teachers, their managers, teacher trainers and learners themselves. This chapter looks at the characteristics – attitudes, knowledge and skills – that are understood to be important in a mathematics teacher. There will be general characteristics that apply to all teachers and some that apply to mathematics teachers. It will be helpful for you to consider your strengths and weaknesses in relation to these characteristics and what you could do to improve any weaknesses you identify.

We also consider a number of elements that may be involved in being a 'professional' teacher of mathematics in the post-compulsory sector after you have completed your initial training.

AN ENTHUSIASM FOR MATHEMATICS AND NUMERACY

Learners often describe how important it is to have a teacher who is enthusiastic about their subject. When you take up teaching as a career, sometimes it can be difficult to be enthusiastic when you are tired and teaching yet another class on percentages, but you really need to be.

What is involved in being enthusiastic? There are a number of things that you do that could suggest enthusiasm. Being aware of mathematics around you will be very helpful. You can find out quite a lot by reading a range of popular literature on mathematics. Authors like Marcus du Sautoy, Ian Stewart, Martin Gardener and Simon Singh write about mathematical ideas and their relationship to the world. There are radio programmes such as the BBC's *More or Less*, which takes a critical look at data in the real world, and Melvin Bragg's *In Our Time*, which, in some episodes, takes a look at a mathematical idea from an historical perspective. There are occasional TV programmes that also look at mathematics – recently Marcus du Sautoy has been involved in a number of programmes that aim to popularise mathematics.

While it is important to be enthusiastic, it is important not to forget that many adults have a negative view of mathematics. Be cautious of not 'overselling'.

GOOD PERSONAL MATHEMATICS SKILLS

One of the key characteristics that has concerned many is the mathematical knowledge of teachers. In the 1980s the Cockcroft report noted these concerns as did the Smith report in 2004. More generally all teachers in the post-compulsory sector have been required to consider a 'minimum core' of literacy, numeracy and information and communications technology (ICT). While they may not teach these subjects explicitly, they are expected to support their learners in developing such skills. In schools, all teachers are expected to undertake tests in English and mathematics before they obtain their professional status.

For specialist teachers there have been a range of programmes to assist subject knowledge, such as the enhancement programmes to assist teachers familiar with the core curriculum to 'upskill' with higher GCSE knowledge. There are also programmes that will help teachers of other subjects to transfer to teaching mathematics.

In addition to direct knowledge of the subject, it is also important for a teacher to see mathematical ideas from a pedagogical viewpoint. For example, you may be happy that dividing fractions involves 'turning the second one upside down and multiplying', but do you have a way of understanding why this happens? Are you aware of the relationship between 'long multiplication' and the distributive law of arithmetic (i.e. $a.(b + c) = a.b + a.c$)? Can you make a link between the various international ways of undertaking division? Such *pedagogical content knowledge* is important to develop – although it will take time.

Throughout this book we have noted how important it is to teach through mathematics in context. In the Netherlands, there is a strand of education called Realistic Mathematics Education (RME) – sometimes Realistic is translated Realisable. In this way of thinking, mathematical concepts are best developed by linking them to some model which can be understood by a learner. Some of these models may be physical ones, like using blocks to develop number – but others can be real contexts that involve mathematics. We have noted in this book many examples of everyday mathematics such as cooking and DIY. If you are to use these effectively then you will need to try them out.

REFLECTIVE SKILLS

An important aspect of teaching is the ability to reflect and develop your practice. There are many models of reflection that will be useful to study later but there are some key elements that can be mentioned here.

As a teacher you will need to think about what you are doing while you are doing it (reflection in practice) as well as looking back after a session (reflection on practice).

You will need to identify positive and negative aspects of your own practice. This can be a difficult thing to do. Some people find it difficult to acknowledge problems in what they do while others can be overly critical and not see the positive aspects. Your mentors and tutors should help you get ideas of what to look for.

When you are reflecting you can do this in many different ways. You may start by considering some strengths and areas for development. Then you may have ideas for why events happened the way they did, and link this to what has been written about teaching. These reflections may suggest changes in practice which can be very helpful. But it does not stop here. You could also consider a wider range of perspectives – perhaps larger sociological or philosophical ideas – which could offer further insights.

In order to reflect upon your practice, you will need to notice what happens with learners. In the end, the purpose of teaching is to assist the learning of the individuals in your group. And how do you recognise learning? It is a difficult thing. Your learners may be able to do the tasks you set – but does that mean they are learning? Perhaps, but perhaps not. It may mean they know how to do that type of task and need to move on to something else. Indeed, you might learn more from the errors that learners make. Part of the job of teaching is understanding what your learners are thinking. The responses to tasks may be documented in various literature or you may have your own ideas of what is in your learners' heads. You will test out your interpretations by asking questions and setting other tasks.

Most of your reflections when you become a teacher will be held in your head. But there are opportunities for recording these on an ongoing basis. Lesson plans will have a space for writing your reflections of the session – these reflections may help in planning further sessions with the group – or for planning future courses. Records of learner progress are another place to record such information for future action. While you are in teacher training, it is expected that you keep a diary of what happens – this can be in a number of forms and might include audio or video recordings.

GOOD COMMUNICATION SKILLS

It is hardly surprising that the communication skills of a teacher are important. A teacher needs to be able to project their voice to a group, to explain ideas and set tasks. You will need to modulate your voice to keep individuals interested and emphasise key elements. You will also need to work effectively on a one-to-one basis and not broadcast an individual's issues to the wider group at times.

You will also need to be able to write appropriately. You will need to write on a board – large writing is a skill that teachers need to develop. Some may already have done so during their education, but it may be new to you. A more recent development is the interactive whiteboard that appears in many institutions and requires another set of skills to work with (more on using ICT later in this chapter).

An important function of a teacher is giving feedback – this could be verbal or written. The 'traditional' way of giving feedback was to note answers to mathematical questions as correct or incorrect. Following work on assessment for learning, it is now thought that feedback should be more constructive and informative. Whether the answer is correct or not, finding out what a learner thinks and identifying correct aspects is important. In addition, it is important to suggest ways forward – again, whether correct or incorrect answers to questions are given there is always a next place to progress.

It is also important to understand what learners are thinking; the ability to ask appropriate questions is significant. There are a whole range of different sorts of questions that could be asked. If you always ask closed questions with one answer, you will be limiting what you understand from learners.

Along with asking appropriate questions, developing skills in listening is a key element to teaching. In a subject like mathematics which can be problematic for learners, it may be even more important to listen to understand how concepts are being developed. It may seem obvious to listen but you do need to remember that a classroom can be a busy environment where it is easy to miss what is said.

Even when you do hear what is said, it is very easy to respond with your own thinking rather than responding to a learner's actual response. One thing to consider trying is a technique favoured by many counsellors – that of 'reflecting back'. Rephrase what the learner has said – this will force you to engage and try to understand the contribution.

AN ABILITY TO ESTABLISH RAPPORT AND TRUST WITH ADULT MATHEMATICS/NUMERACY LEARNERS

We have already noted that mathematics and numeracy learners come with a range of motivations, prior experiences and views about learning. Some of these learners – at all levels – are quite likely to have anxieties about learning mathematics. Even the most successful individuals can have doubts and concerns about their next steps.

Recognising this is a good starting point for working with learners. There may be times where you ask someone to do something that they feel uncomfortable with – if you are able to notice this happening then you can think of ways of reducing the discomfort.

It is worth noting that you will need to understand the boundaries of relationships with learners. The institution will have policies around the extent to which relationships should develop and will probably depend upon age. For example, there may be a requirement to not meet learners off site, to not share personal contact details. You may also have your own personal boundaries. Nevertheless, whatever boundaries exist it will be important to gain trust and build rapport in order to be able to support.

FLEXIBILITY AND OPENNESS TO DIFFERENT APPROACHES

Whatever your own views of teaching and learning, you are most likely to find that you will be asked to develop courses and run sessions in ways that are not consistent with your position. This pressure could be from learners, from colleagues, from managers or from external institutions such as those requiring inspections (Her Majesty's Inspectorate [HMI] and Office for Standards in Education [Ofsted]) or those proposing innovation (for example the National Centre for Excellence in the Teaching of Mathematics).

A professional will need to recognise that while a teacher has a certain amount of autonomy, they also act as a representative for their organisation. As such they will not necessarily have the freedom to choose exactly what they want to do. At times, it may be important to present ideas counter to the consensus – and, indeed, to do so vigorously – but also you will need to recognise that if you wish to work within an organisation you may need to rethink your own position.

This is not to suggest that you should always be compliant; it is important to have debates about teaching and learning. Indeed, it is possible that your own views may become part of a future consensus. On the other hand, we have also seen many individuals change their views of learning – and that would not happen if they were not open to different perspectives.

Alongside the need to be flexible and open to ideas is the willingness to act upon these new views. Teachers are usually very busy individuals, planning and preparing, delivering and assessing groups of learners. But there is also a need to save some time and effort for self-development. This could be via courses or short events, or it could be your own research through reading texts or online investigations. This self-development is usually known as continuing professional development (CPD) and most professions involve some form of updating in this way. We return to professional development beyond initial teacher training later.

Preparation for classes can take a fair amount of time. Like most activities, the more experienced you are, the quicker you can do this. But, there is not enough time to prepare resources for all of your sessions. Having said this, using 'off the peg' resources has its limitations. They will not necessarily have the context you want, or use the numbers you want. And the majority of resources are fairly standard worksheets which, while useful, should not be the only type of resource used.

We have shown a number of different examples of resources throughout this book. One of the things you can do is take a worksheet and convert the examples into an entirely different resource.

REFLECTIVE AND RESPONSIVE TO FEEDBACK

Throughout your practice as a teacher you will receive feedback from learners, colleagues, managers and, potentially, other external individuals. We have already noted how important reflection is as a skill for teachers. You will need to reflect on this feedback from a range of interested individuals.

We have already noted the importance of reacting positively to comments from others. It is important to consider feedback carefully. You may not agree with all of it but you should consider all feedback in order to change your future practice.

GOOD SKILLS IN THE USE OF INFORMATION AND COMMUNICATIONS TECHNOLOGY

It is fairly obvious to note that using information and communications technology (ICT) is important in the modern world. This is true for all teachers who will be using ICT both in the classroom, preparing for teaching and as a fundamental part of their administrative and record keeping procedures. But mathematics also brings some specific issues of its own.

》 *Producing documents*: probably using word processing packages but also other software such as spreadsheets and presentation software. Using these software packages does present some issues for the mathematics teacher that might not be so important for others.

How do you write fractions? Do you use an oblique (for example, 4/5)? Do you use specialist software like mathtype (for example, $\frac{4}{5}$) which produces an embedded object in Word? Do you use the table function?

$$\frac{4}{5}$$

Each of these has their advantages and disadvantages; for example, the tables are easily editable but do not fit neatly into a paragraph of text.

There are specialist symbols which you might need to use depending upon the level, for example.

Even the four operation signs may be an issue. While the + sign is straightforward, the hyphen (-) and the subtract (−) signs are a little different in size (the subtract sign is the hyphen with symbol font), the letter X and the multiplication sign × are different and, of course, the division sign ÷ is a specialist object. None of these are difficult to find but do need some exploration on the computer.

❯ *Using specialist software and websites*: there are a number of dedicated sites and pieces of software that can be used to support mathematical development. It is worth investigating the range of resources that exist on websites such as BBC Skillswise, NIACE, NCETM and the National STEM Centre, among others.

There are some interesting interactive activities that can be used but you need to decide whether these would be displayed as a plenary to a group or whether to use individual or paired use.

- You might use computers for learners to research some ideas or data which the can be used with the whole group.

- Interactive whiteboard software usually includes some mathematics specific resources such as graph paper, grids and shapes, which can be useful in teaching.

- Using general software with learners – not only can you produce word-processed documents and spreadsheets for your own use but you can also use specifically designed documents with learners. The drop-down selection tool within most modern word processors can be used as a way to leave an optional choice. A spreadsheet can also use 'gap fill' tasks which can then automatically indicate whether the selection works or not. For example, you can set text to change to green when a correct response has been input or red for an incorrect response.

- As a professional it is expected that you will engage in record-keeping and administrative duties. This will involve a range of ICT skills which could include accessing databases and using spreadsheets.

 TASK 8.1

Consider the list of characteristics and identify your strengths and weaknesses. How could you work on your weaknesses?

Suggested responses to tasks can be found at the end of the book.

BEYOND INITIAL TEACHER TRAINING AND BEING A PROFESSIONAL

So what does it mean to be a professional? There is no simple, straightforward answer to this.

One issue is around qualifications. Teaching qualifications have been the normal requirement for teaching in schools for a long time, whereas in the post-compulsory sector this has not been the case. The requirement was only introduced by the Labour Government in 2000 with specialist adult numeracy training following shortly after. More recently, the Coalition Government removed the requirements (at least for most teachers, although they retained a requirement for adult numeracy). Despite the removal of the requirement, many institutions still ask for a qualification and many teachers prefer to be trained and have a qualification on their CV.

Another recent development was the introduction in 2007 of the Qualified Teacher Learning and Skills (QTLS) status. The idea was that a professional working in the post-compulsory sector would join a professional body – called the Institute for Learning (IfL) – and evidence their professional status in much the same way as other professionals such as accountants and doctors. Teachers were required to evidence a minimum of 30 hours of professional development each year to hold on to their status as teachers within the sector. QTLS was designed to be a 'licence to practise'.

Along with the removal of the requirement for a teaching qualification, was the removal of the requirement to join the IfL and gain the status of QTLS. Both the organisation and the status still exist and are being used by many practitioners. Again, for some it is seen as a positive on a CV when applying for jobs.

But these are all structural aspects of professionalism that were introduced to mirror other professions. The term professionalism can be used to cover other aspects of the teaching role.

The characteristics that we have mentioned above are part of the answer. They are aspects that prepare you to be a professional but then these characteristics need to be acted upon. After you have gained your teaching qualification, you may not be surprised to learn there is plenty more development to undergo. In the same way that doctors, lawyers and accountants are expected to keep their practice up to date, so should a teacher, even if not required to record such developments with a professional body.

Such ongoing professional development will involve the types of aspects noted in the task above. There is subject knowledge enhancement – which may involve some new knowledge around already known mathematics as well as new ideas, and may involve the discussion of pedagogical issues, ideas for teaching and the development of a range of resources.

The term 'lifelong learning' has been introduced to recognise that all individuals should consider themselves as on a learning journey. The teacher of mathematics in the post-compulsory sector is one of those individuals who will need to continually evolve in knowledge and practice.

FINAL THOUGHTS

We leave you with a few questions to ponder on the issue of professionalism. It is not set as a task, as the issues have already been raised.

❱ What do you think being a professional teacher of mathematics in the post-compulsory sector should involve?

❱ Should all teachers be required to obtain a teaching qualification? What are the advantages and disadvantages of such a requirement?

❱ Should there be a professional body for teachers in the post-compulsory sector that everyone should join? What are the advantages and disadvantages?

❱ Should you be required to evidence your professional development? Again, what are the advantages and disadvantages of doing this?

❱ Is mathematics and/or numeracy different to other subjects with respect to professionalism?

FURTHER READING

DfES (2005) *Improving Learning in Mathematics*. Sheffield: Department for Education and Skills Standards Unit.

Education and Training Foundation (2014) *2014 Professional Standards for Teachers and Trainers*. At: http://www.et-foundation.co.uk/wp content/uploads/2014/05/4991-Prof-standards-A4_4-2.pdf

Websites

CURRICULA

The **Adult Numeracy Core Curricula** is found on the Excellence Gateway
http://www.excellencegateway.org.uk/sflcurriculum

Functional mathematics criteria, Ofqual criteria
http://ofqual.gov.uk/documents/functional-skills-criteria-for-mathematics/
(More information at: http://www.excellencegateway.org.uk/node/20516)

GCSE Mathematics, Ofqual criteria
http://ofqual.gov.uk/documents/gcse-subject-criteria-for-mathematics/

Free-standing mathematics qualifications developed through Nuffield
http://www.nuffieldfoundation.org/fsmqs

AS/A Level mathematics criteria, Ofqual criteria
http://ofqual.gov.uk/documents/gce-as-and-a-level-subject-criteria-for-mathematics/

Finally, **core mathematics** in development at the time of writing
https://www.gov.uk/government/publications/core-maths-qualifications-technical-guidance

RESOURCES

The following sites have resources that can be used for teaching mathematics and numeracy.

BBC Skillswise has a number of videos and paper resources.
http://www.bbc.co.uk/skillswise/

Skills Workshop has a range of user-generated resources.
http://www.skillsworkshop.org/

The **Excellence Gateway** is a site for the post-compulsory sector that contains resources from a range of projects. The Skills for Life and Embedded Skills resource sets are particularly useful.
http://www.excellencegateway.org.uk/node/18239

The **National Centre for Excellence in the Teaching of Mathematics** has a range of resources for schools and the post-compulsory sector.
https://www.ncetm.org.uk/resources/

The **National STEM Centre** has a library of resources including mathematics.
http://www.nationalstemcentre.org.uk/elibrary/

ONLINE COURSES

Learning Mathematics Online is produced by NIACE with Learning Unlimited.
http://learningmathsonline.ac.uk/

Citizen Maths is funded by the Ufi Charitable Trust. It is developed by Calderdale College, with CogBooks, the Institute of Education and OCR, with advice from the Google Course Builder team. http://citizenmaths.com/

ORGANISATIONS

Membership organisations

Adults Learning Mathematics – A Research Forum (ALM) is an international research and practitioner organisation (conference proceedings and journal articles are freely available online).
http://www.alm-online.net

The **National Association for Numeracy and Mathematics in Colleges (NANAMIC)** is a UK-based practitioner society for the post-compulsory sector.
http://www.nanamic.org.uk

Research and Practice in Adult Literacy (RaPAL) is primarily focused on literacy but also includes numeracy in its brief.
http://rapal.org.uk

National organisations

The **National Institute for Adult Continuing Education (NIACE)**, a membership body and registered charity, is a long-established development organisation and think tank that works on adult education of all kinds.
http://www.niace.org.uk

The **National Research and Development Centre for adult literacy and numeracy (NRDC)** was established in 2002 with a focus on adult literacy, numeracy and ESOL.
http://www.nrdc.org.uk

National Numeracy was set up in 2012 and is a charity formed to work on mathematics/numeracy across all phases.
http://www.nationalnumeracy.org.uk

Key texts

We have provided some further reading in each chapter but thought it might also be helpful to have the following key texts identified here.

Casey, H. *et al.* (2006) *You Wouldn't Expect a Maths Teacher to Teach Plastering… Embedding Literacy, Language and Numeracy in Post-16 Vocational Programmes: The Impact on Learning and Achievement.* London: National Research and Development Centre for adult literacy and numeracy (NRDC). At:
http://nrdc.org.uk/publications_details.asp?ID=73# [accessed 1 September 2014].

Coben, D. *et al.* (2007) *Effective Teaching and Learning: Numeracy.* London: National Research and Development Centre for adult literacy and numeracy (NRDC). At:
http://nrdc.org.uk/publications_details.asp?ID=86# [accessed 1 September 2014].

Griffiths, G. and Stone, R. (eds) (2013) *Teaching Adult Numeracy: Principles and Practice.* Maidenhead: Open University Press.

Hodgen, J., Coben, D. and Rhodes, V. (2009) *Feedback, Talk and Engaging With Learners: Formative Assessment in Adult Numeracy Research Report.*

London: NRDC/King's College London. At:
http://www.nrdc.org.uk/publications_details.asp?ID=180# [accessed 1 September 2014].

NRDC (2007) *Thinking Through Mathematics: Strategies for Teaching and Learning.* London: National Research and Development Centre for adult literacy and numeracy (NRDC). At:
https://www.ncetm.org.uk/online-cpd-modules/ttm/

Parson, S. and Bynner, J. (2006) *Does Numeracy Matter More?* London: National

Research and Development Centre for adult literacy and numeracy (NRDC). At:
http://nrdc.org.uk/publications_details.asp?ID=16# [accessed 1 September 2014].

Swain J., Baker, E., Holder, D., Newmarch, B. and Coben, D. (2005) *'Beyond the Daily Application': Making Numeracy Meaningful to Adult Learners.* London: National Research and Development Centre for adult literacy and numeracy (NRDC). At:
http://nrdc.org.uk/publications_details.asp?ID=29# [accessed 1 September 2014].

Swan, M. (2006) *Collaborative Learning in Mathematics: A Challenge to Our Beliefs and Practices.* Leicester: NIACE and NRDC.

Suggested answers to tasks

1. THE LEARNERS

TASK 1.1

a) The literature usually talks of a number of 'personal and social' factors affecting learning. These are often broken down into social factors (gender, age, ethnicity), experiences (employment and interests), language and literacy, cognitive (preferred ways of learning, learning difficulties) and affective (motivations) subheadings.

b)

(i) Previous experiences can be positive and negative. It is worthwhile noting that such experiences often create expectations of what being in a class is about. These expectations may require a period of transition when trying new types of learning activities.

(ii) It can be very important to harness learners' knowledge and use it effectively within the classroom. We encourage teachers to explore learners' experiences and generate class discussion.

(iii) Learning difficulties may affect memory requiring extra time and the use of memory aids. Ideas for teaching those with learning difficulties are raised in Chapter 7.

(iv) If you use a lot of learning activities which the learner feels are unsuited then you may find the learner reluctant to attend sessions. While the idea of fixed 'learning styles' has been challenged recently, it is important to listen to learners and to consider how to adapt activities.

(v) Learners who come from a war zone are likely to suffer from a number of psychological issues. They may need extra time for thinking, or indeed time for themselves. Do discuss with the individual what they feel they need.

(vi) Learners have busy lives and may be distracted at times. Be aware of this and discuss issues with learners.

(vii) The individual ICT skills of learners are important, and in particular access to technology may affect a learner's ability to be involved in various activities.

(viii) Similar to (ii) above.

(ix) Again there will be similar issues to (ii) and (viii) above, but time constraints are also likely to be an issue.

(x) Most learners in classes will have taken part in compulsory schooling, but there will be individuals who may have had limited access to education and need much more introduction to topics.

(xi) See (vi) but also be aware that the learner may want to know about the way that mathematics is taught in schools and a teacher may need to research this.

(xii) Previous learning experiences can bring back memories for learners. You may need to try different types of activities – again, talking to learners is important.

2. MATHEMATICAL KNOWLEDGE

TASK 2.1

Suggested answers

Binary	Denary
110	6 (4 + 2)
1000	8 (8 + 0 + 0 + 0)
1010	10 (8 + 2)
11011	27 (16 + 8 + 0 + 2 + 1)
100010	34 (32 + 2)

Note that 1011 bin, 100110 bin, 12 den and 33 den do not have matches.

TASK 2.2

Suggested answers

Binary calculations

101011 + 111 = **110010**

1	0	1	0	1	1
			1	1	1
1	1	0	0	1	0
	1	1	1	1	

1011 – 110 = **101**

		1	^{1}0	1	1
			1	1	0
			1	0	1

$111011 \times 1101 = \mathbf{1011111111}$

				1	1	1	0	1	1
						1	1	0	1
				1	1	1	0	1	1
			0	0	0	0	0	0	0
		1	1	1	0	1	1		
	1	1	1	0	1	1			
1	0	1	1	1	1	1	1	1	1
1	1	1	1	1	1				

$11011 \div 1001 = \mathbf{11}$

							1	1
1	0	0	1	1	1	0	1	1
				1	0	0	1	
				0	1	0	0	1

TASK 2.3

Suggested answer

For example, it is important that the parts in a motor vehicle are precisely engineered, whereas buying paint to decorate a room probably cannot be accurate because different walls absorb paint differently. When administering drugs in intensive care for the young or old, it is important to be accurate, whereas deciding how much drink is needed for a party need not be so precise.

Interestingly, the NRDC report *Measurement* (2007) noted that, 'Many learners do not identify measure as part of mathematics, despite using measure more often than many other areas of mathematics. Many adult learners grew up using the imperial system of measure and now have to use the metric system. This creates confusion both in the units they use and in the way they read the measuring instruments they choose to use.' One might argue that measurement is the point of mathematics. If we want to build tunnels, develop medicines and send objects into space, then we need to measure and calculate the amounts of various quantities. Measurement also requires that we understand fractions and decimals. Importantly, the NRDC report argued that the curriculum should be focused on measure rather than emphasising decontextualised number.

TASK 2.4

In education, there is some terminology – sometimes known as meta-language – that is used to describe some of these different ways of calculating. In the following different ways of calculating, note the use of the following terms: partitioning, bridging and chunking.

Addition

Calculation	Answer	Possible method
2 + 8	10	**Number bonds** – know what adds up to 10
4 + 8	12	**Bridging** – know that 4 + 6 adds to 10 and then 2 more are left to be added
12 + 10	22	Understand adding 10s
17 + 15	32	Use number line from 17, 3 more to 20, 2 more to 22 and 10 more to 32

18 + 19	37	**Near doubles**: double 18 is 36 add 1
28 + 46	74	Partitioning 20 + 40 = 60 8 + 6 = 14 Together make 74
125 + 8	133	**Bridging** – 125 add 5 is 130, add remaining 3
145 + 19	164	Compensation 145 + 20 = 165, 165 − 1 = 164

Subtraction

Calculation	Answer	Possible method
10 − 7	3	Number bonds
24 − 8	16	Bridging 24 − 4 = 20 20 − 4 = 16
33 − 20	13	Subtracting tens
23 − 14	9	Adding on/number line. 14 to 20 is 6, 20 to 23 is 3 6 + 3 = 9
75 − 23	52	Partitioning 70 − 20 = 50 5 − 3 = 2 50 + 2 = 52
54 − 6	48	Bridging 54 − 4 = 50, then take away 2 more = 48
54 − 19	35	Compensation 54 − 20 = 34 34 + 1 = 35

Multiplication

Calculation	Answer	Possible method
3 × 4	12	Use multiplication fact/times table
8 × 4	32	Use fact or double twice 8 × 2 = 16 16 × 2 = 32
12 × 10	120	Multiplying by ten/hundred, etc. means changing the place value

12 × 15	180	Use 15 = 10 + 5 12 × 10 = 120 12 × 5 = 60 (half of the line above) 120 + 60 = 180
18 × 19	342	**Compensation** 19 is 1 less than 20 18 × 20 = 18 × 2 × 10 = 360 So 18 × 19 = 360 − 18 = 342
28 × 23	644	**Partitioning** 23 = 20 + 3 28 × 20 = 560 28 × 3 = 84 560 + 84 = 644

Division

Calculation	Answer	Possible method
12 ÷ 3	4	Use multiplication facts 3 × 4 = 12 means 12 ÷ 3 = 4
24 ÷ 6	4	Use multiplication facts 4 × 6 = 24, or Dividing by 6 is the same as dividing by 2 and then by 3 (6 = 2 × 3) 24 ÷ 2 = 12, 12 ÷ 3 = 4
70 ÷ 10	7	Dividing by ten, means changing the place value: 7 tens becomes 7 units.
90 ÷ 15	6	**Repeated subtraction** 90 − 15 = 75 75 − 15 = 60 60 − 15 = 45 } 6 lots 45 − 15 = 30 30 − 15 = 15 15 − 15 = 0
90 ÷ 6	15	'Chunking': repeated subtraction in chunks 10 × 6 = 60 so break 90 into 60 + 30 60 ÷ 6 = 10 and

		$30 - 6 = 24$ $24 - 6 = 18$ $18 - 6 = 12$ } 5 lots $12 - 6 = 6$ $6 - 6 = 0$ $10 + 5 = 15$
$240 \div 12$	20	$10 \times 12 = 120$ $240 = 120 + 120$ So $240 \div 12$ $= 120 \div 12 + 120 \div 12$ $= 10 + 10 = 20$

TASK 2.5

On a scale of 0 to 1, with 0 as impossible and 1 as certain, you will have probably put a) close to zero and c) close to 1. b) would depend upon the time of the year, d) is actually quite high (depending upon when the cancer is discovered) and (e) and (f) are much harder to determine.

TASK 2.6

1. Only includes the readership of that paper and only those who respond.
2. Only includes children of that school and won't include adults without school-age children.
3. Only includes those with a telephone and also there will be more than one person for most telephone numbers.

3. EDUCATIONAL THEORY AND APPROACHES TO TEACHING MATHEMATICS AND NUMERACY

TASK 3.1

There are certainly aspects of mathematics that require psychomotor skills. The ability to use equipment to measure is important in mathematics. Being able to position a ruler or a protractor would be a psychomotor skill. The ability to write the symbols is another. What is probably harder to see is how the affective domain relates to mathematics. It is perhaps no surprise to know that mathematics provokes feelings within individuals. Some people find mathematics a wonderful problem-solving tool while others can be filled with dread at the thought of the subject. But these are feelings about mathematics. Do feelings play a part in mathematics itself? Perhaps a view of what counts as an appropriate answer to a question might form part of this. Indeed, for some mathematicians aesthetics are involved in mathematics – an 'elegant solution' is often noted. This is part of the affective domain, although where this fits in to post-compulsory mathematics is a question left open for the reader to consider.

TASK 3.2

There are a number of aspects of fractions that could be considered.

You could look to develop an understanding of the meaning of fractions with objectives such as

- ❱ 'Identify one amount as a fraction of another' which might involve activities that match fractions with two amounts.
- ❱ 'Describe the meaning of fraction' which might involve asking learners to complete a partial definition.

You could be aiming for learners to calculate a fraction of an amount with objectives such as:

- ❱ 'Calculate a fraction of a given amount' which might involve a 'follow me' card set (see page 74) involving fractions of a single amount such as £360.
- ❱ 'Identify how to calculate fractions of an amount' which might involve selecting an appropriate calculation from a given list.

You could develop thinking around fraction equivalencies with objectives such as:

- ❱ 'Identify equivalent fractions' which might involve a memory matching game.
- ❱ 'Convert a fraction into a percentage' which might involve a jigsaw game with matching fractions and percentages.

TASK 3.3

The first thing to note is that learners are all different and it is important to listen to what they say. This means building in time to your course to talk to learners about their backgrounds, goals and views.

If someone has experienced failure in mathematics in their past, it is quite likely that this will be repeated if the same approaches from the past are used. It may be a good idea to try different approaches and to encourage learners to share ideas of how they understand the subject.

Some learners come along not to gain a qualification but to help them in their lives. For such an individual, preparing for external examinations may not be the most significant aspect of education. Rather, it will be important to have mathematical examples placed in a variety of contexts. Some of the most common include shopping, cooking, budgeting, DIY and sports. These may also be helpful for qualification-focused courses, although more or less time can be used in understanding the ideas rather than examination preparation.

In the case of the learner who wants to help their children there are some things to consider which would not be the case for other learners. Such individuals would want to know the mathematics delivered in schools and may also need ideas to help them support their children.

There may be other individuals who want to understand more about the world around them and see study as part of a general development.

While these groups of individuals may benefit from some similar activities, it is important to bear in mind their different motivations and find out about their backgrounds in order to plan accordingly.

TASK 3.4

The birth rates are likely to prompt discussion of what is an 'appropriate' birth rate and immigration issues. These may be problematic areas, particularly if not thought about in advance. At the same time, there are potentially important conversations that could be had about religion, immigration and other matters of interest to all. It may allow for a real discussion of equal opportunities and appropriate language for classroom discussions.

4. THE TEACHING AND LEARNING CYCLE

TASK 4.1

Suggested answers

The four items a)–d) are intended to deal with the same area of mathematics but raise different issues. We do not intend to provide an answer as to which approach is 'best' – because that is not possible – but we can discuss the advantages and disadvantages of each item. We suggest that you evaluate a wide range of different items and strategies when considering assessment processes.

a) and b) are fairly standard ways to ask questions. a) is a fairly pure number example, whereas b) is in a context. There are pros and cons to each; answering b) incorrectly may be related to language issues, whereas for some the decontextualised nature causes its own problems.

c) is intended to minimise the stress caused to the individual and to require less time. It is also a measure of the learner's self confidence in understanding percentages. The problem is that without trying the question it may be difficult for some to answer.

d) allows individuals to write what they know about working out percentages. The problem is that some or many individuals will be unsure of what to write and if language and literacy skills are not strong this may be particularly difficult – especially if the task is a written one.

TASK 4.2

There are no simple right or wrong answers to this and to some extent it depends upon who inputs the information on the plan – you or the learner. If these were proposed by a learner you might want to check what aspects are really their priorities. For instance, does a) mean learning multiplication tables or having some strategy for multiplying larger numbers? b) is more specific than 'convert between units' as it specifies metric lengths, but are there specific measures the learner is concerned with? c) might mean drawing bar charts or it might mean interpreting what a bar chart means. It is the job of the teacher as the professional with knowledge of the curriculum to discuss and suggest the details.

TASK 4.3

There are a number of possibilities and it depends upon the level of the group and what you are intending to develop. For a) you could ask 'How can you make 72?' which would allow for all operations; you could encourage them to try all four operations and find as many different ways as possible or estimate the answer. By doing this you will allow all learners to contribute. For b) you could ask 'What do you know about millimetres?' which could allow for a range of responses. For c) you could ask 'What do you know about pentagons?' or perhaps 'A triangle has three sides; how many sides do other shapes have?' A follow-up question might be 'Are all five-sided shapes pentagons?'

TASK 4.4

There is no specific response to this task but you could include the following points:

❭ Encourage self-motivation

❭ Share group knowledge

❭ Encourage learners to express their thoughts

TASK 4.5

The responses here can give the teacher some information which then can be followed up. Response a) may indicate some confusion between addition and subtraction. This could be conceptual – individuals may not know how to subtract such numbers and do what they are able to do – or a problem with interpreting the instruction due to dyslexia or partial sightedness. Of course it could just be a misread. b) does suggest some sense of subtraction and large numbers; the problem is that there are too many nines involved. This might be a simple error but does suggest that an individual needs some kind checking procedures to look at answers. c) is the correct answer, but as is the case with correct answers you may still want to test thinking to find out whether it can extend to other answers (what about subtracting 2 or 11?). d) does evidence some aspect of subtraction but does suggest some problem with either the algorithm used or with checking mechanisms (or both).

TASK 4.6

Look at the example lesson plans for ways of differentiating.

Here are some points to consider when planning. Think about the individuals in the class you are planning to teach. Think about those that you expect to find the topic straightforward and those who will find it challenging. What is the best way of supporting the learner who is struggling while finding further challenge for the learner who is further ahead in their understanding?

5. TEACHING PRACTICE

TASK 5.1

Here are some sample activities.

Materials required	Topic	Description of activity
String, scissors, tape measures	Estimation of metric length and area. Accurate measurement using a tape measure. Calculation of area.	Hand out sections of string to small groups of learners. Ask learners to estimate how long a metre is and mark the string accordingly. Follow with a discussion on how decisions were made. Learners use a tape measure to accurately measure a metre and cut string to correct length. Next, place four lengths together to form a square. Learners estimate how many squares would cover the floor of the room. Follow up with a discussion on what the square represents and what they have estimated. Learners estimate dimensions of room and use estimates to get an approximation for floor area. Learners use tape measure to measure dimensions and calculate area accurately.
Post-it notes, coloured pens	Collecting and organising data	Choose a topic such as method of transport to class or day of the week that people were born on (small groups of learners could choose different topics and ask the rest of the class to record their data). Individual learners record their data on the post-it note and attach it to a wall or other surface. Discuss how to organise the data so that it is easier to read; for example, post-it notes may be stacked in columns by day of the week to form a bar chart. Ask learners to create labels for the chart using post-it notes.

TASK 5.2

Suggested answer

a) What learners might expect of a teacher.

- To be prepared and punctual
- To be knowledgeable and enthusiastic about the subject
- To prepare the learners for relevant accreditation
- To be fair and consistent
- To mark set work within an agreed timescale if appropriate
- To give clear and constructive feedback on work and on progress

b) What learners might expect of each other.

- To be punctual
- To respect other learners and their opinions

- To listen when the teacher and other learners are talking

- To switch off mobile phones during the session (or have them on silent)

Task 5.3

Suggested answer

Possible advantages of being in a training class:

❱ Receiving moral support from your peers – safety in numbers.

❱ Getting ideas from each other for planning teaching activities and resources.

❱ Observing your peers trying out different approaches.

❱ Getting to know individual learners while someone else is teaching.

❱ A gentle introduction to group teaching – only having to plan and deliver teaching for short periods.

❱ Getting different viewpoints on your teaching by receiving feedback from several people.

Possible advantages of being in a placement:

❱ Receiving one-to-one support from your mentor.

❱ Getting stuck into teaching more quickly and gaining direct teaching experience at a faster pace.

❱ You may feel more involved and accepted as a class teacher rather than a visitor.

Possible disadvantages of being in a training class:

❱ Having to share the class teacher or mentor with several other people – the extra demands on the teacher's time may make it difficult to get information or feedback from them.

❱ If the class is small or attendance low, it may feel like there are too many of 'you' in the room and you may not get an opportunity to support individuals or small groups.

❱ Personality clashes could occur within the training class group which may make joint planning or feedback sessions difficult.

❱ Someone else might get to teach the topic that you were hoping to teach.

❱ You may feel like a visitor rather than a class teacher.

Possible disadvantages of being in a placement:

❱ You may feel that you are asked to teach before you feel ready.

❱ You will only get to observe one particular teaching style – i.e. that of the class teacher.

6. EXAMPLES OF TEACHING ACTIVITIES

TASK 6.1

Starter activities can be devised for all areas of mathematics. Examples can be found within the resources found online (see the websites section). In addition to these sites, a number of school-based sites that have useful interactive activities may be used (although be aware that some learners can get put off by items that look 'childish'), for example http://www.transum.org.

TASK 6.2

At the start of Chapter 3 we make the point that the relationship between theory and practice is not a straightforward one. If we take the example of the money 'investigation', the activity can be used as a collaborative one and involves an important cultural product (in this case money) which can be interpreted through humanist as well as cognitive perspectives. The measurement activity involves dealing with psychomotor as well as cognitive skills but also uses real objects in an attempt to engage with everyday, social (i.e. cultural) activity. The IT activities are intended to reflect the increase in the distribution and use of IT in the world around us. The language activities are intended to address multicultural and 'ethnomathematical' approaches but also provide an opportunity for collaborative learning. Games are collaborative activities and can allow for discussion of significant mathematical ideas; they also can link to social and cultural aspects of people's lives. It may be important to ask for participants to help devise games from their own cultures.

Many of the activities in this chapter could be discussed from a 'critical' perspective, to argue about their significance to participants and in the real world from their perspectives.

7. PLANNING AND INCLUSIVE PRACTICE

TASK 7.1

Measurement is an activity that occurs in many environments. The following are suggested examples in everyday and employment contexts. They are intended to be of interest to a range of individuals – there are certainly plenty more examples that could be chosen in each case.

Everyday contexts:

- Shopping
- Cooking
- DIY
- Budgeting

Employment contexts:

- Hairdressing
- Architect
- Builder
- Caterer

TASK 7.2

A range of approaches to teaching can be found in *Access for All* which can be accessed online. It is important to not assume too much about given conditions and to realise that you will need to discuss issues with individual learners. In addition, individuals will bring strengths with them. Some dyslexic learners may be good at holistic approaches to problem solving, seeing the 'bigger picture' and being able to estimate or interpret an answer.

Reading measuring instruments might be a challenge to dyslexic learners if the markings are small – this may also be true of the partially sighted.

When using activities that require learners to move around the room it will be important to discuss them with individuals. These might include activities such as arranging people in order of their height to exemplify the median, or carrying out measuring activities around the room.

Drawing graphs may present difficulties for a range of learners including those with dyspraxia.

Learners on the autistic spectrum find it difficult to understand idioms – for example, you might say 'I wonder if you could ...' and intend this as an instruction but it may not be received as such. Writing down clear rules in simple sentences can help such learners.

TASK 7.3

Note: There are alternatives to the answers given.

a) Mode b) Modal	A pair of words that refer to the same mathematical concept but differ in grammatical use.
c) Fifth	A word for a number that also shows relative position (order).
d) Find	A common word used differently in the context of numeracy and maths problems.
e) Work out	A common phrase used differently in the context of numeracy and maths problems.
f) Kezban	An unrecognisable word that may cause confusion (in this case a name).
g) Possible	A word with a precise meaning in statistics but a less specific meaning in everyday life.
h) Risk	A word that has a particular negative connotation in everyday life.
i) Circle	A common description of a shape that may require defining in mathematics.
j) Football	An activity used to provide context that may not be familiar to some – or connote an alternative such as American football (culturally specific).
k) Cooking	An activity used to provide context that may not be familiar to some (culturally specific).
l) Dinner	An everyday event that may be perceived differently according social class.

8. PROFESSIONALISM

TASK 8.1

There is clearly no direct way to respond to this very personal task. But there are some actions you could consider undertaking. These include the following.

CPD of various forms, such as:

- subject enhancement
- ideas for teaching
- materials development
- listening skills
- voice training
- writing skills

Your own professional development, for example:

- observing another teacher
- reading, listening, viewing a range of media
- taking part in online courses

Appendix

Exemplars of planning

Extract of a scheme of work from a Level 1/2 Functional Mathematics course

Topic		Learning objectives *By the end of the session the learners should be able to:*
Week Date:	Decimal calculations Long multiplication Long division	Explore the different methods of long multiplication: grid, lattice and traditional. Explore the different methods of division: traditional and chunking.
Week Date:	Currency exchange Multiplying and dividing decimals	Convert exchange rates from sterling into foreign currencies: decimal multiplication. Convert from foreign exchange rates to sterling: decimal division.
Week Date:	Fractions	Find parts of whole quantities using fractions. Use fractions to add and subtract amounts or quantities. Evaluate one number as a fraction of another.
Week Date:	Percentages	Calculate 1% and 10% of a value. Calculate VAT at 20%. Calculate percentage increases/decreases. Find one number as a percentage of another.

Extract of a scheme of work from a discrete numeracy course, Entry 3–Level 2

SCHEME OF WORK

Course/Subject: Numeracy

Session No.	Content	Outcome *At the end of the session learners will be able to:*	Assessment techniques	Resources
1	Course induction and assessment Assessment and course induction Place value and ordering	Familiarise themselves with the content of the course and each other Identify strengths and areas for development with respect to numeracy skills Identify place value from millions to hundredths	Observation of activity Self-assessment activity Written assessment Questions	Course guide Assessment pack Icebreaker sheet Place value grids Playing cards
2	Number – whole number Set ILP targets Strategies for doing calculations Rounding	Agree ILP targets Carry out calculations using efficient methods Approximate whole numbers by rounding	Discussion Feedback on activities Questions	ILPs A3 paper Multiplication square IWB Handouts
3	Measure, shape and space Length, area and perimeter	Identify appropriate units of length to use Estimate and measure length Calculate area and perimeter Convert between different units of length	Questions and feedback Discussions Observe activities Check written work	Rulers, tape measures Squared paper String Activities Worksheets
4	2D and 3D shapes Shape vocabulary Recognition and properties of shapes	Recognise and name common 2D and 3D shapes Sort 2D and 3D shapes according to their properties Solve problems using mathematical properties of regular 2D shapes	Observation of activities Feedback on activities Questions	2D and 3D shapes Shape name cards Scissors Mirror tiles Handouts IWB

Example lesson plan 1

Title	Family Numeracy – Understanding Shapes		
Tutor(s) Week		Date Time Location	9:30 – 12:00

Aim of session: To equip learners to support their children with the topic of shape

			Assessment of objectives
Learning objectives. The learners will be able to: Name, recognise and describe 2D and 3D shapes and their properties (MSS2/E2.1; E2.2; L1.1) Identify the difference between 2D (flat) and 3D (solid) shapes (MSS2/E3.1) Identify lines of symmetry (MSS2/E3.1; L1.1) Sort shapes according to their properties (MSS2/E3.1) Identify right angles in a room (MSS2/E2.4; L1.1) *Make a resource to take home to use with their children* *ECM: Enjoy and achieve*			Observation Discussion and questions Completed activities Completed task sheets

Time	Topic/Activity	Resources
9:30	**Introduction of the topic.** **Discuss shapes in everyday life:** Invite learners to find and name shapes in the room. Ask learners if they know what 2D and 3D mean.	
9:40	**Sorting and matching 2D and 3D shape names:** Small-group activity. Tutor hands out shape name cards to each group and they place it with 2D or 3D headings. Brief discussion on unfamiliar shape names. Tutor checks results then asks whole group what shapes are called. Groups review their results. Groups then match shapes with names. **Differentiation**: Group learners by level. Entry 2 work on basic shapes; Entry 3/Level 1 to include rhombus, trapezium, parallelogram, triangular and hexagonal prism, tetrahedron.	Shape name cards, 2D and 3D shapes

Time	Activity	Resources
10:20	**Symmetry:** Tutor elicits understanding of symmetry and explains the meaning of lines of symmetry. Working in small groups, learners try to record total number of lines of symmetry for the given 2D shapes. **Differentiation:** Open-ended task. Learners working at Entry 2 expected to identify symmetry; Entry 3 expected to identify line of symmetry; Level 1 expected to identify number of lines of symmetry in range of shapes.	Task sheet, scissors, mirrors
10:50	**Right angles:** Learners cut out angle munchers and tutor demonstrates how to use them. Learners use them to look for right angles in the room. Discussion on angles. Tutor hands out school sheet on shapes and invites learners to ask questions about things they are unfamiliar with. **Differentiation:** Using support materials and differentiated task. Entry 2 look for right angles in room using angle muncher. Entry 3 look for right angles in photographs and sort according to number of right angles. Level 1 look for right angles in photographs and construct right-angled shapes.	Angle munchers, scissors School information sheet on shape Task sheets Set squares Protractors
11:10	Break	
11:20	**Shape properties and vocabulary:** Tutor shows a shape and asks learners to describe it without saying the name. Tutor writes property vocabulary on the board. **Differentiation:** Targeted questioning. Target learners working at higher levels when showing more complex shapes.	Shapes, IWB
11:35	**Materials making:** Making a shape Pelmanism game. Tutor explains how game works then supports learners with making the game. If time, learners try out the game.	Pelmanism sheet, card, coloured pens, scissors
11:50	**Evaluation:** Ask for feedback from learners and to complete ILPs. Hand out shape grid for homework and test papers (at different levels) as required.	ILPs, shape grid sheet, test papers

Example lesson plan 2

Title	Negative numbers	
Date:		
Aims of session: To reinforce skills associated with negative numbers		
Learning objectives. **The learners will be able to:** Find the difference between positive and negative numbers Compare positive and negative numbers using the inequality symbol Add and subtract positive and negative numbers Use a calculator to carry out calculations involving positive and negative numbers		**Assessment of objectives** Observation of activities Feedback from activities Discussion and questions Peer assessment Self-assessment Written exercises Homework

Time	Topic	Teacher and learner activity	Resources
12.00	Learning objectives	Tutor shares learning objectives and talks through them and activities. Tutor completes register.	Register PowerPoint
12.05	Starter – temperature	Tutor asks learners when they use negative numbers in their lives. Tutor explains activity to learners. Hands out cards to small groups of learners and asks them to match them. Tutor asks learners if they were surprised by any of the answers.

Differentiation: Aim for support through groups. Tutor provides support where needed. Learners self-assess. | PowerPoint Matching cards, answer sheets |

Time	Topic	Activity	Resources
12.15	Negative numbers on a number line	Tutor goes through strategy using PowerPoint. Asks targeted questions to check for understanding. Tutor then facilitates whole group activity – hands out mini whiteboards to learners. One learner to write down a positive number and another learner to write down a negative number. Other learners to calculate the difference. Repeat as required for understanding. Tutor goes through some of the questions on altitude with whole group – asking questions. Learners respond using mini whiteboards. **Differentiation:** Encourage peer support. Target learners for assessment. Tutor provides support where needed. Learners given the choice of using a number line.	PowerPoint Mini whiteboards and pens Number lines Text books
12.30	Finding the difference using a number line	Individual work on finding the difference – using textbook. **Differentiation:** Tutor gives support as necessary and guides learners towards appropriate level of questions.	Number lines Textbooks
12.50	Break		
1.00	Inequalities	Tutor introduces learners to inequalities and assesses prior knowledge through questioning. Tutor hands out negative digit cards and inequality signs to pairs and introduces activity. Learners work in pairs – each person selects a card and pairs agree on which way to place the sign. **Differentiation:** Tutor notes prior knowledge/experience. Use mixed level pairings for activity. Tutor provides support as necessary.	Negative digit cards Inequality signs
1.10	Inequalities	Tutor introduces problem involving inequalities. Pairs work on problem and feed back on how they approached it. Tutor leads discussion on feedback and highlights any misconceptions. Learners try inequalities exercise from textbook. **Differentiation:** Mixed-level pairings. Option of peer support and use of number line.	PowerPoint Text book Number lines

1.25	Adding and subtracting + and – numbers game	Tutor sets out, explains and demonstrates game and puts learners in teams. Learners play game. Tutor facilitates and supports with scoring. **Differentiation:** Peer support from teams. Tutor gives support where needed.	Number cards, coin, dice, calculators Hand out on adding and subtracting positive and negative numbers
1.40	Adding and subtracting + and – numbers	Learners work individually on exercises from textbook. **Differentiation:** Tutor gives support where needed. Provide higher-level worksheet where appropriate.	
1.55	Summary	Tutor sums up and asks for feedback. Tutor gives out homework information and informs learners about next session. **Differentiation:** Give learners who are interested higher-level questions.	Homework: Exam questions – pg. 93 Higher-level questions

Notes:
For mixed pairings/groupings, pair/group any of H, F, BT with any of BL, AL, AV.
Higher-level work for H and F (possibly BT)

Index